The Detection of Septicemia

Editor

John A. Washington II
Professor of Microbiology and of Laboratory Medicine
Mayo Medical School
Head, Section of Clinical Microbiology
Mayo Clinic
Rochester, Minnesota

Published by

CRC PRESS, Inc.
2255 Palm Beach Lakes Blvd. · West Palm Beach, Florida 33409

Library of Congress Cataloging in Publication Data

Main entry under title:

The Detection of septicemia.

Bibliography: p.
Includes index.
1. Septicemia–Diagnosis. 2. Bacteriology, Medical –Cultures and culture media. I. Washington, John A.
RC182.S4D39 616.9'44'075 77-12652
ISBN 0-8493-5207-X

This book represents information obtained from authentic and highly regarded sources. Reprinted material is quoted with permission, and sources are indicated. A wide variety of references are listed. Every reasonable effort has been made to give reliable data and information; but the author and the publisher cannot assume responsibility for the validity of all materials or for the consequences of their use.

This monograph was written by the author(s) in his (their) private capacity. No official support or endorsement by the U.S. Environmental Protection Agency or any other agency of the federal government is intended or should be inferred. Mention of commercial products does not constitute endorsement by the U.S. Environmental Protection Agency.

International Standard Book Number 0-8493-5207-X

Library of Congress Card Number 77-12652

Printed in the United States

PREFACE

Septicemia represents a medical emergency that requires the prompt institution of therapy. The choice of antibiotics is usually empirically based upon several bits of information among which are the patient's underlying disease, antecedent procedures or instrumentation, the presence and site of preexisting infection, prior antimicrobial therapy, the endemicity or epidemicity of certain microorganisms in the hospital, and the antimicrobial susceptibility of these microorganisms. The laboratory's role is to isolate and identify the etiologic agent as rapidly as possible and to determine its susceptibility to various antimicrobial agents.

This rather complex sequence of events requires close interaction between those involved directly in the patient's care and those involved in the microbiological examination of the patient's specimens. The sooner the microbiologist can isolate the etiologic agent and determine its susceptibility to antibiotics, the sooner the clinician can render more specific antimicrobial therapy in appropriate dosages.

Our objectives in preparing this monograph are to define septicemia and the factors contributing to its occurrence, to describe its incidence at the Mayo Clinic and affiliated hospitals, to review in detail the variables that influence the recovery of organisms from cultures of blood, and to describe several newer approaches to the detection of bacteremia in the laboratory.

Although much has been done in recent years to study many of the factors that are important in cultures of blood, it will be readily apparent to the careful reader of this monograph that in certain areas or topics our information reflects more the state of the art than the state of the science. Much, therefore, remains to be done to fill in the gaps in our knowledge. As new information becomes available, new questions will be posed and new recommendations, often contrary to existing ones, will be made.

The reader is forewarned that some of the data presented in this monograph may seem contradictory and even confusing, reflecting to a great degree the enormous diversity of blood culture systems in use against which other systems have been compared. Because of the varieties of media and their specified and unspecified additives, as well as the differences in their preparation, bottling, atmospheres of incubation, examination, and subcultures, it is often difficult to draw broad conclusions from the literature. It is, therefore, essential to scrutinize the details about the systems used and to assess their adequacies and inadequacies, as well as their advantages and disadvantages, relative to other systems.

I am grateful to my coauthors for their participation in this monograph: Walter R. Wilson, M.D., of the Division of Infectious Diseases, Duane M. Ilstrup, M.S., of the Section of Medical Research Statistics, and John P. Anhalt, Ph.D., M.D., of the Section of Clinical Microbiology. Acknowledgment is certainly due to Ed Warren, Administrative Coordinator of the Section of Clinical Microbiology, Marsha Hall, Supervisor of the Bacteriology Laboratory, and the many technologists in the Section for their participation in and support of the studies we have conducted for the past decade. Finally, I wish to thank Lola Jaeger for her able assistance in the preparation of this manuscript.

John A. Washington II, M.D.
Rochester, Minnesota

THE EDITOR

John A. Washington II, M.D. is Head of the Section of Clinical Microbiology at the Mayo Clinic, Rochester, Minnesota and Professor of Microbiology and of Laboratory Medicine at the Mayo Medical School. Dr. Washington earned his B.A. degree from the University of Virginia in 1957 and received his M.D. degree from the Johns Hopkins University in 1961.

Prior to joining the staff of the Mayo Clinic, Dr. Washington was a house officer in General Surgery at Duke University Medical Center, a Clinical Associate at the National Cancer Institute, and a resident in Clinical Pathology at the National Institutes of Health.

Dr. Washington is a Fellow of the American Society of Clinical Pathologists, the American College of Physicians, and the American Academy of Microbiology. He is a member of the Infectious Diseases Society of America and the American Society for Microbiology. He is currently a member of the editorial boards of the *American Journal of Clinical Pathology* and *Antimicrobial Agents and Chemotherapy.* He has published over 100 scientific articles on various aspects of clinical microbiology and the activity of antimicrobial agents, has edited a book entitled *Laboratory Procedures in Clinical Microbiology,* and is Associate Editor of the 16th edition of *Clinical Diagnosis by Laboratory Methods.*

CONTRIBUTORS

John P. Anhalt, Ph.D., M.D.
Consultant
Section of Clinical Microbology
Mayo Clinic
Assitant Professor of Microbiology
and Laboratory Medicine
Mayo Medical School
Rochester, Minnesota

Duane M. Ilstrup, M.S.
Statistician
Section of Medical Research Statistics
Mayo Clinic
Rochester, Minnesota

John A. Washington II, M.D.
Head of the Section of Clinical
Microbiology
Mayo Clinic
Rochester, Minnesota

Walter R. Wilson, M.D.
Consultant
Division of Infectious Diseases
Mayo Clinic
Assistant Professor of Medicine
Mayo Medical School
Rochester, Minnesota

TABLE OF CONTENTS

Chapter 1

SEPSIS

DEFINITIONS, UNDERLYING CONDTIONS, MANIFESTATIONS

Walter R. Wilson

The terms bacteremia and sepsis are frequently used interchangeably. In this book, bacteremia is defined as blood cultures containing bacteria or fungi, while sepsis, or septicemia, is defined as blood cultures containing bacteria or fungi in association with clinical and laboratory signs of infection, i.e., fever, chills, tachycardia, hypotension or shock, leukocytosis, etc.

UNDERLYING CONDITIONS PREDISPOSING TO SEPTICEMIA

Bacteremia in Normal, Healthy Adults

The possibility that healthy individuals may experience transient, asymptomatic bacteremia has been the source of controversy since Reith and Squier[1] reported in 1932 that 12% of 99 apparently healthy persons had positive blood cultures; however, the majority of the organisms isolated in their study may have been contaminants. MacGregor and Beaty[2] reported that 152 of 1707 (8.9%) blood cultures were judged to contain contaminants. In a recent study using current skin antisepsis, blood collection and inoculation procedures, blood culture bottles and subculture techniques, 2.1% of 240 patients who had no demonstrable foci of infection had positive blood cultures.[3] In each of these patients, the organism isolated probably represented contamination rather than bacteremia. These reports were similar to the presumed contamination rate of 1.9% (166 of 8654 cultures) reported in a previous study of detection of bacteremia.[4] These studies suggest that true bacteremia occurs rarely, if ever, in a healthy individual in the absence of manipulative procedures.

Manipulative Procedures

Bacteremia has been reported to occur after a variety of manipulative procedures: rocking the teeth with forceps,[5] tooth extraction,[6] urologic procedures,[7-18] proctosigmoidoscopy,[19] colonoscopy,[20-22] liver biopsy,[23,24] nasotracheal suction,[25] esophageal dilatation,[26] barium enema,[27] fiberoptic bronchoscopy,[28] angiography,[29] and cardiac catheterization.[30] Bacteremia associated with these procedures is usually self-limited. The patients may be asymptomatic or may experience transient fever, chills, and other signs of infection; however, in patients with preexisting valvular heart disease, infective endocarditis may result.

Sepsis associated with manipulative procedures occurs most frequently following genitourinary tract manipulations in patients with preexisting urinary tract infection.[18] In these patients, complications include hypotension; septic shock; osteomyelitis, especially of the lumbar vertebrae; infective endocarditis; and meningitis.[31]

Venipuncture and the Use of Intravenous Therapy (Table 1)

The risk of transmission of infection during venipuncture was emphasized in an early report by Mendelssohn and Witts,[32] which noted that direct inoculation of skin microflora into peripheral veins during venipuncture may occur. Katz et al.[33] stated that back flow of blood may occur from evacuated blood collection systems and may result in bacteremia. McLeish et al.[34] reported five temporally related cases of *Serratia* bacteremia and found that 3 of 13 evacuated blood collection tubes contained *Serratia.* In a recent study by Washington,[35] 9% of reportedly sterile evacuated blood collection tubes contained bacteria including *Pseudomonas aeruginosa, Serratia marcescens,* enterococci, and *Acinetobacter calcoaceticus.*

Intravenous infusion is an indispensable form of medical therapy, but intravenous therapy-associated septicemia is a potential life-threatening complication. When percutaneously inserted plastic catheters were left in place for longer than 48 hr, the associated septicemia rate was reported to be from 2 to 5%[36-45] and may be as high as 8%.[46,47] These observations have resulted in the recommendation that plastic intravenous catheters be replaced after 48 hr.[36] The skin microflora is

TABLE 1

Sepsis – Underlying Conditions and Common Microbiologic Cause

Condition	Microorganisms
Venipuncture and intravenous therapy	
Venipuncture (contaminated systems)	*Serratia*
Infected catheter tips	*S. epidermidis, S. aureus, Klebsiella, Enterobacter, Serratia,* enterococci, *Candida*
Contaminated intravenous fluids	*Enterobacter, Klebsiella*
Intravenous hyperalimentation	*Candida*
Contaminated blood or blood products	*Enterobacter, Salmonella*
Respiratory	*Pseudomonas, Serratia, S. aureus,* other Gram-negative bacilli
Gastrointestinal	
Bowel	*B. fragilis, E. coli, Enterobacter Klebsiella,* group D streptococci, anaerobic cocci, clostridia
Hepatobiliary	*E. coli, Klebsiella, Enterobacter,* group D streptococci, *Serratia*
Spontaneous peritonitis	*E. coli, S. pneumoniae*
Cardiovascular system	
Endocarditis	Viridans streptococci, group D streptococci, staphylococci, Gram-negative bacilli
Infected vascular grafts	*S. epidermidis, S. aureus E. coli, Klebsiella, Serratia*
Genitourinary tract	
Urinary tract infections	*E. coli, Klebsiella, Proteus,* enterococci, *Pseudomonas*
Female pelvic infections	*B. fragilis,* peptostreptococci, clostridia, *N. gonorrhoeae,* group B streptococci
Immunosuppressed patient	*Pseudomonas aeruginosa, E. coli, S. aureus, Candida*

the most likely source of microorganisms isolated from catheter tips.[48–57] Those most commonly isolated have been *Staphylococcus epidermidis* and *Staphylococcus aureus, Klebsiella, Enterobacter, Serratia,* enterococci, and *Candida.* It has been hypothesized that skin microorganisms gain access to the tip of the catheter at the time of insertion and subsequently by migrating along the interface between catheter and tissue. Microorganisms from distal sites may also infect the catheter tip. The loosely organized clot that forms quickly around the intravascular segment of the plastic catheter could serve as a trap and culture medium for circulating microorganisms. Transient bacteremias from distant sites may seed the clot surrounding the catheter tip. *Candida* infection of catheter tips may occur in patients receiving long-term broad-spectrum antimicrobials.[58–62] In these patients, the skin microflora is altered, thereby favoring the overgrowth of *Candida,* which might gain access to the catheter tip through the site of insertion.

The use of iodine-containing disinfectants applied to the skin around the catheter insertion site is effective in reducing the frequency of catheter-associated infections.[63–68] The value of topical ointments containing bacitracin, neomycin, and polymyxin in the prevention of catheter-associated sepsis is less clear.[39,40,43,48] The use of antibiotic ointments may increase the frequency of colonization of catheter tips by *Candida* and antibiotic-resistant bacteria.[39,40,43] A potentially fatal complication of intravenous therapy is suppurative phlebitis. *S. aureus* and multiply drug-resistant *Klebsiella-Enterobacter* have been implicated most frequently in this infection.[69–79]

Contamination of intravenous fluids may cause sepsis. In 1970, major outbreaks of nosocomial sepsis were related to contamination of elastomer-lined screw-cap enclosures of intravenous fluid bottles.[36,80–86] *Enterobacter* and *Klebsiella* have been the most frequent isolates from patients with sepsis caused by contaminated i.v. fluids.[87–89] *Enterobacter* species and other members of the tribe *Klebsielleae* proliferate rapidly at room temperature in commercial solutions of 5% dextrose and water, reaching concentrations of $> 10^5$ organisms per milliliter within 24 hr.[81–84]

Sepsis may occur in association with the use of intravenous hyperalimentation. The frequency of sepsis has ranged from 0 to 27% of patients.[90–106] In a survey conducted by the Center for Disease Control, Atlanta, Ga., of 31 hospitals treating 2078 patients with intravenous hyperalimentation, sepsis occurred in 7% of patients; 54% of these cases were caused by *Candida.*[98] This is in contrast to the relative rarity of fungal sepsis in patients receiving conventional intravenous infusion therapy.[81–85,107–130] A number of factors have been proposed which may predispose patients who are receiving hyperalimentation to sepsis caused by *Candida.* These patients are usually chronically ill, debilitated, and immunosuppressed and are frequently receiving long-term parenteral, broad-spectrum antimicrobials that predispose them to fungemia.[58–62] In addition, the use of topical antibiotic ointments applied to the skin catheter insertion site may favor the overgrowth of *Candida.* Moreover, *Candida* grows well in fluids used for intravenous hyperalimentation. Infection control measures suggested by Dudrick, et al.[90] and Maki and associates[36] have considerably reduced the frequency of sepsis associated with the use of parenteral hyperalimentation.

Contaminated blood and blood products have been reported to cause septicemia.[131–135] One to 6% of individual donor units have been found to be contaminated with bacteria.[136–138] In most cases, sepsis was caused by contamination with microorganisms capable of growth at 4°C. (psychrophilic bacteria) such as *Enterobacter* and *Salmonella choleraesuis.*[139]

Respiratory Source

In recent years, the frequency of sepsis originating from hospital-acquired pulmonary infections caused by Gram-negative bacilli has increased.[140–152] This is primarily the result of the increased use of ventilatory assistance devices in chronically ill, immunosuppressed patients. Some Gram-negative bacilli, such as *Pseudomonas* and *Serratia,* grow well in water, chemical disinfectants, and other solutions used in mechanical ventilatory equipment.[144,147,153–155] The contamination of reservoir nebulizers has resulted in the direct aerosolization of Gram-negative bacilli in the respiratory tract. Patients with neurologic diseases, alcoholics, or chronically ill, obtunded patients are predisposed to aspiration pneumonitis which may result in sepsis caused by Gram-negative bacilli or anaerobic microorganisms.[156–159] *S. aureus* is part of the commensal microflora of the nasopharynx, and *S. aureus* septicemia may originate from this area, particularly in immunosuppressed patients.[143]

Gastrointestinal Tract

Sepsis originating from the gastrointestinal tract is usually related to previous abdominal surgery, hepatobiliary disease, malignancy, inflammatory bowel disease, or diverticulitis.[160–171] The microorganisms most frequently isolated from patients with these disorders are *Bacteroides fragilis;* Enterobacteriaceae, especially *E. coli, Enterobacter,* and *Klebsiella;* group D streptococci; and anaerobic streptococci. Polymicrobial bacteremia with mixed aerobic and anaerobic Gram-negative bacilli and Gram-positive cocci is often associated with an intra-abdominal focus.[172,173]

Intra-abdominal infections are the most common portal of entry for Gram-negative anaerobic sepsis and *B. fragilis* accounts for 58 to 90% of the anaerobic Gram-negative bacilli recovered from blood cultures.[173–178] The typical patient with *B. fragilis* bacteremia is a man between 30 and 60 years old with some underlying gastrointestinal disorder (frequently carcinoma of the colon) in whom an infected surgical wound or intra-abdominal complication developed 10 days postoperatively.[173] An interesting association exists between *Clostridium septicum* bacteremia and carcinoma of the colon;[179] clinicians should suspect an occult intra-abdominal malignancy in patients with bacteremia caused by this microorganism without an obvious portal of entry.

Sepsis, in association with hepatobiliary disease, is most often caused by *E. coli, Klebsiella, Enterobacter,* group D streptococci, and *Serratia.*[145,146,160,180] Biliary obstruction, cholelithiasis, cholecystitis, and cholangitis or empyema of the gall bladder are the most frequent precipitating causes. Spontaneous peritonitis and sepsis caused by *Streptococcus pneumoniae* or *E. coli* may occur in patients with cirrhosis.[181-183] Impaired function of the hepatic reticuloendothelial system and the presence of circulating inhibitors of polymorphonuclear leukocyte chemotaxis[183-185] have been postulated as factors predisposing patients with cirrhosis to spontaneous peritonitis.

Cardiovascular System

The use of intravenous therapy is the most common cause of sepsis originating from the vascular system, while the use of permanent intravascular prostheses is another major source. Patients with infected cardiac valve prostheses may be divided into two groups – early-onset infections (<2 months postoperatively) and late-onset infections (>2 months postoperatively).[186-188] Among patients with early-onset infections, sepsis is most likely caused by bacterial contamination of prosthetic valves during the intraoperative period.[189,190] *S. aureus* and "opportunistic" microorganisms such as *S. epidermidis*, Gram-negative bacilli, and corynebacteria are the most frequent isolates from these patients.[186-188, 191-194] In patients with late-onset prosthetic valve endocarditis, viridans streptococci, Gram-negative bacilli, *S. aureus*, and *S. epidermidis* are the most frequent isolates.[186-188,191-198] Infection in these patients is thought to be most often the result of transient, clinically inapparent bacteremia.[199]

Insertion of intravascular grafts,[200-215] transvenous pacemakers,[216-221] and ventriculocardiac or ventriculoperitoneal shunts[222-238] may cause sepsis, the cure for which is usually not possible without removal of the infected prosthesis. Infection may occur as a consequence of intraoperative contamination, spread from a contiguous area of infection, or, rarely, secondary to bacteremia from a distant focus of infection. *S. epidermidis, S. aureus, E. coli, Klebsiella, Serratia*, and other Gram-negative bacilli are the most common causes of infection.

Genitourinary Tract

Approximately 50% of patients with aerobic Gram-negative bacillary sepsis have associated urinary tract infections as the suspected portal of entry.[146,239-241] Studies have shown that the use of an indwelling urinary bladder catheter for 3 days or longer is associated with an 80 to 100% risk of development of urinary tract infection.[146,242] As many as 10% of patients with catheter-associated urinary tract infections develop bacteremia.[146] Bacteremia has been reported to occur in association with transurethral resection of the prostate (31% of patients), cystoscopy (17%), urethral dilatation (25%), and urethral catheterization (8%).[18] The majority, but not all, of these patients had preexisting urinary tract infections. *E. coli*, enterococci, *Proteus, Klebsiella*, and *Pseudomonas* are the most frequent urinary tract isolates. Not uncommonly, these infections are hospital-acquired, and the microorganisms may be multiply drug-resistant.

Postpartum and postabortion infections, endometritis and pyometria, and gonococcal pelvic inflammatory disease are causes of sepsis in females,[243-245] with *B. fragilis*, peptostreptococci, clostridia, and *Neisseria gonorrhoeae* the most frequent isolates. Clostridial sepsis is of particular interest. Historically, clostridial infections have been considered among the most dreaded complications of postabortal sepsis, war wounds, and other traumatic injuries. Tetanus, gas gangrene, and overwhelming infections with hemolytic anemia, renal failure, and shock are clinical entities that justify this onerous reputation. However, patients with clostridial bacteremia often have a benign course and not uncommonly recover without specific antimicrobial therapy.[174] Ramsay[246] reported that 17 of 28 gynecologic patients with *C. perfringens* bacteremia had localized infections without systemic toxicity. Bacteremia due to *Clostridium ramosum* also has been reported to have a favorable prognosis despite lack of specific antimicrobial therapy.[247]

Immunosuppressed Patient

In the immunosuppressed patient, neutropenia, cytotoxic, radiation, immunosuppressive therapy, and abnormal T- and B-cell functions associated with underlying disease processes render the host exquisitely susceptible to lethal infections caused by virtually any microorganism. Infections are frequently caused by microorganisms which are relatively avirulent and which possess only limited capacity to produce invasive infection in normal hosts. Pneumonia and perirectal infections with resultant septicemia are among the most frequent life-threatening infections in immunosuppressed hosts.[144,251,254] The frequency of episodes of pneumonia is related to the state or activity of the

TABLE 2

Clinical Manifestations Suggestive of Gram-Negative Bacteremia

Chills, fever and hypotension
Fever alone (particularly in patients with malignancies, hematologic, urinary tract or gastrointestinal disorders, and following instrumentation or insertion of urinary or intravenous catheters)
Hyperpnea, tachypnea and respiratory alkalosis
Oliguria or anuria
Hypotension
Thrombocytopenia
Change in mentation (agitation, confusion, stupor)
Acidosis
Hypothermia
Evidence of urinary tract infection
} Without apparent cause
Evidence of pulmonary infection (associated with ventilatory assistance)
Ecthyma gangrenosum

From McCabe, W. R., Gram-Negative Bacteremia, in Dowling, H. F. (Ed.), *Disease-A-Month,* December, 1973. Copyright © 1973 by Year Book Medical Publishers, Inc., Chicago. Used by permission.

underlying disease. Leukemic patients who relapse have a much higher frequency of fatal pneumonias and septicemias than do leukemic patients who are in remission. *P. aeruginosa, S. aureus,* and *E. coli* are the most frequent bacterial causes of septicemia in immunosuppressed patients, while septicemia caused by anaerobic microorganisms is surprisingly uncommon. In the immunosuppressed host, *P. aeruginosa* is the most frequent cause of sepsis originating from a perirectal source, but *Listeria monocytogenes* is an important cause of sepsis in renal transplant patients and other immunosuppressed patients.[253,254] Splenectomized patients are predisposed to fulminant, rapidly fatal sepsis caused by *S. pneumoniae.*[255,256]

MANIFESTATIONS

Clinical Features

The clinical manifestations of sepsis range from transient asymptomatic bacteremia to fulminant infections with bacteremic shock and death occurring within a few hours. Irrespective of the bacterial etiologic agent, the features of sepsis are similar, but the severity of infection and prognosis of patients with sepsis are usually dependent upon patient age; underlying associated conditions, such as leukemia, carcinoma, chronic obstructive pulmonary disease; and other poorly understood factors. Classically, the patient with sepsis exhibits shaking chills, sweats, fever, prostration, and, occasionally, nausea and vomiting (Table 2). The patient's skin may be warm or cold, depending upon whether vasodilatation or vasoconstriction has occurred. In severe cases, hypotension followed by oliguria or anuria and septic shock may appear, usually within 4 to 10 hr after the initial manifestations.[239,249–265] The classic sequence – chills, fever, prostration, hypotension – is easily recognizable clinically. However, only about 30 to 40% of patients with sepsis demonstrate this typical pattern.

Clinical manifestations of sepsis are more subtle in the majority of patients. Fever is almost always present at the onset of sepsis; it may be minimal or absent in patients receiving corticosteroids, salicylates, or other antipyretic therapy, in the elderly, and in patients with uremia. In patients who are afebrile or lack other classic signs of sepsis, unexplained confusion, agitation, stupor, hypotension, tachypnea, or respiratory alkalosis may be the initial manifestation of sepsis. Rarely, sepsis may present with hypothermia.[146]

Cutaneous skin lesions, usually located on the distal extremities, are developed by 70 to 90% of patients with gonococcal sepsis.[261,262] Typically, the lesions begin as an erythematous tender maculopapule, become vesicular, then evolve rapidly to an ulcerative lesion with a shaggy grey necrotic center surrounded by an erythematous base. The appearance of tender indurated or ulcerated cutaneous lesions with black necrotic centers – ecthyma gangrenosum – is characteristic

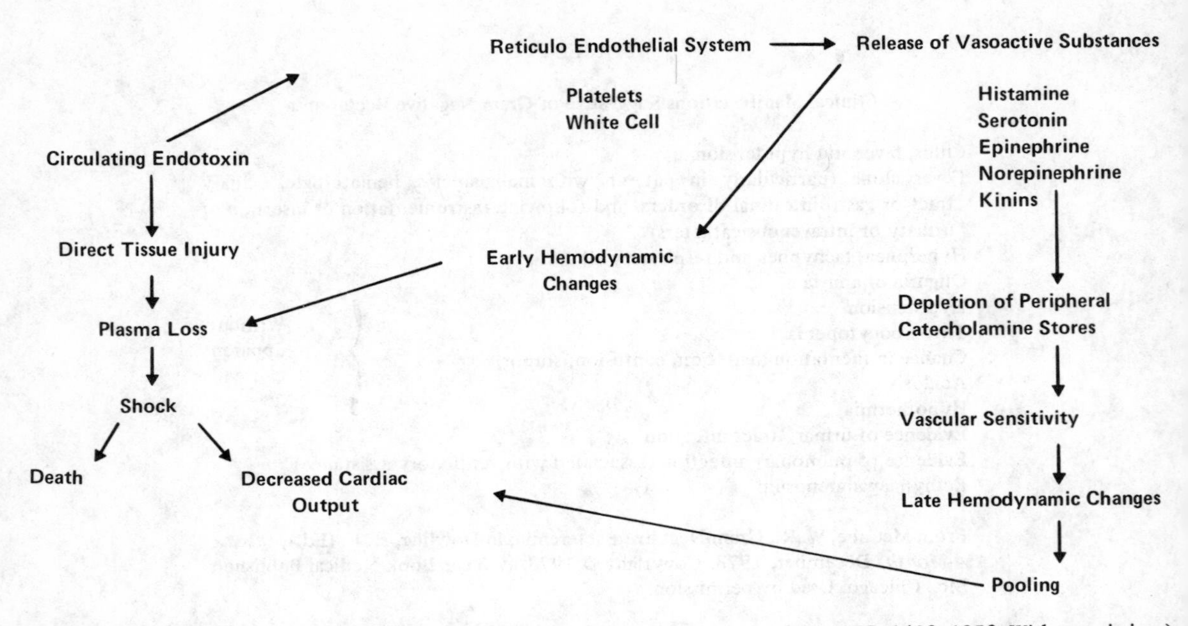

FIGURE 1. Effect of circulating endotoxin. (From Hassen, A., *Med. Clin. North Am.*, 57, 1403, 1973. With permission.)

of *Pseudomonas* sepsis.[146,263] Similar lesions may appear with sepsis caused by *Aeromonas* and *Citrobacter.*[146,264] *S. aureus, B. fragilis,* and *Nocardia asteroides* sepsis may be associated with subcutaneous abscesses;[173,174,265] maculopapular skin lesions have been reported in patients with *Candida* sepsis.[266] Extensive ecchymotic and cutaneous hemorrhages associated with intravascular coagulation and fibrinolysis may occur in association with severe sepsis caused by any microorganism, but occur typically in patients with meningococcal, pneumococcal, and rickettsial sepsis.[267-269]

Metastatic abscesses in the liver, brain, kidney, bone, lung, and other tissues may occur in patients with sepsis, especially in association with infections caused by *S. aureus, Bacteroides, Nocardia,* and anaerobic streptococci.[173,174,266] Septic arthritis occurring in patients with sepsis is most frequently caused by *N. gonorrhoeae, S. aureus,* and *S. pneumoniae.*[261,270] Acute bacterial osteomyelitis is most frequently seen in pediatric patients and results from *S. aureus* septicemia.[271] Thrombophlebitis and jaundice have been reported to occur in association with *B. fragilis* septicemia.[173,174]

Septic shock has been defined variously by several investigators[239,257-260] but is generally described as sustained hypotension and clinical signs of decreased organ perfusion, i.e., oliguria, confusion, and peripheral vasoconstriction occurring in association with bacteremia. Septic shock occurs more frequently with Gram-negative bacillary bacteremia than with Gram-positive coccal bacteremia. A significant decrease in blood pressure occurs in approximately 40% of patients with Gram-negative bacillary bacteremia. Hypotension usually appears shortly (within 4 to 10 hr) after the initial manifestation of bacteremia. The pathophysiologic events leading to septic shock are complex, interdependent mechanisms triggered by endotoxemia and are schematically represented in Figure 1. Often the first sign of septic shock is alteration in the patient's mental status, most commonly confusion and disorientation. As the severity of shock increases, the patient may become somnolent or semicomatose and may have pale, damp, cold skin, collapsed jugular and other peripheral veins, and a rapid, feeble arterial pulse. The distal extremities may become cyanotic. Tachypnea and shallow respirations with or without signs of pulmonary congestion develop. Tachycardia is common, but bradycardia may occur in elderly patients. In severe cases, progressive oligo-anuria develops, circulatory collapse with lactic acidemia and severe acidosis occurs, and death usually follows.

Pulmonary abnormalities occur frequently in association with septic shock caused by Gram-

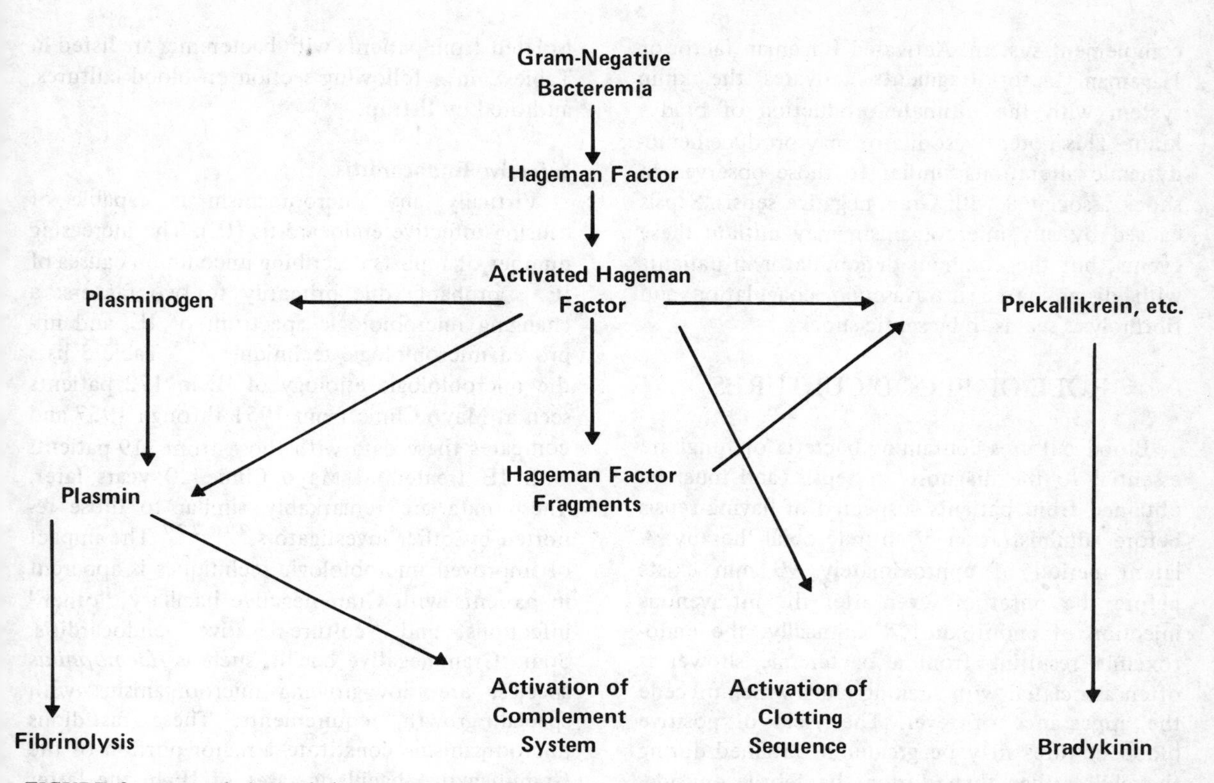

FIGURE 2. Interrelations of changes in the coagulation, fibrinolytic, complement, and kinin systems in Gram-negative bacteremia. (From McCabe, W. R., in Dowling, H. F. (Ed.), *Disease-A-Month,* December, 1973. Copyright © 1973 by Year Book Medical Publishers, Inc., Chicago. Used by permission.)

negative bacilli and have been termed "shock lung."[272-274] Physiologically, impaired ventilation – perfusion – and alveolar hypoventilation evolve with resultant hypoxemia. Chest X-rays in these patients resemble bacterial, viral, or fungal pneumonitis or acute pulmonary edema. Histologically, alveolar thickening and pulmonary edema take place.

Laboratory Abnormalities

Leukocytosis is a common feature of sepsis. Endotoxin enhances the adhesive properties of granulocytes, causing them to marginate on endothelial surfaces, which results in an initial brief leukopenia followed by a leukocytosis. Endotoxin may also have a direct stimulatory effect on granulopoiesis. Occasionally in elderly patients and commonly in immunosuppressed patients, Gram-negative sepsis may be associated with leukopenia. Thrombocytopenia occurs in 50 to 60% of patients with Gram-negative bacteremia.[146-269] Other laboratory abnormalities may include increasing hyperazotemia, hyperkalemia, hypoglycemia, and hypoferremia. Tachypnea may produce respiratory alkalosis. In septic shock, profound hypoxemia, metabolic and respiratory acidosis, and lactic acidemia may develop.

While thrombocytopenia occurs commonly in patients with Gram-negative bacillary sepsis, disseminated intravascular coagulation and fibrinolysis are uncommon – <5% of patients – and are limited almost exclusively to patients with septic shock.[146,269] In these patients, platelets, and coagulation factors II, IV, VIII, and fibrinogen are characteristically low, and serum levels of fibrin split products are elevated. These abnormalities result from complex alterations and interactions among the coagulation, fibrinolytic, kinen, and complement systems schematically represented in Figure 2. The initial precipitating event is probably activation of Hageman factor (factor XII) which subsequently results in activation of the intrinsic clotting system and the fibrinolytic system. Activation of the complement cascade system which interacts with the clotting system may occur via the classic C1423 pathway as a result of antigen-antibody combination or by direct activation of C′3 by endotoxin. Plasmin may also activate the complement system. Activated Hageman factor or Hageman factor fragments activates the kinin system with the ultimate production of bradykinin. This potent vasodilator may produce hemodynamic alterations similar to those observed in shock associated with Gram-negative sepsis. Sepsis caused by any microorganism may initiate these events, but the common denominator in patients with disseminated intravascular coagulation and fibrinolysis seems to be septic shock.

ROLE OF BLOOD CULTURES

Blood cultures containing bacteria or fungi are essential to the diagnosis of sepsis, and must be obtained from patients suspected of having sepsis before administration of antimicrobial therapy. A latent period of approximately 90 min exists before the onset of fever after the intravenous injection of endotoxin.[146] Clinically, the endotoxemia resulting from a bacteremic shower is often associated with shaking chills which precede the appearance of fever. The yield of positive blood cultures may be greater if obtained during the chill rather than during the febrile episode. However, more important than the timing of venipuncture for blood culture is the use of appropriate blood culture media and adequate volumes of blood for culture. Moreover, depending on the clinical setting and index of suspicion for a specific microbiologic etiology, selective blood cultures should be obtained for fungi, *Brucella,* and *Neisseria* in addition to routine blood cultures.

The identification of an isolate from blood cultures frequently provides a clue as to the source or portal of entry of sepsis. Conversely, if the source of sepsis is known, knowledge of the predisposing conditions frequently provides the clinician with a high index of suspicion for the putative microbiologic agent. These factors are most important in the selection of appropriate initial antimicrobial therapy and provide valuable guidelines for additional therapeutic measures such as incision and drainage of abscess, removal of infected intravenous cannulae, vigorous bronchopulmonary physiotherapy, etc.

ORGANISM INVOLVED

The organisms isolated from patients with bacteremia, seen at Mayo Clinic from 1968 to 1975, are listed in Table 1, and the organisms isolated from patients with bacteremia are listed in Table 2 in a following section on blood cultures, authored by Ilstrup.

Infective Endocarditis

Virtually any microorganism is capable of causing infective endocarditis (IE). The increasing number of reports describing uncommon causes of IE is probably due primarily to two factors: a changing microbiologic spectrum of IE, and improved microbiologic techniques.[275] Table 3 lists the microbiologic etiology of IE in 172 patients seen at Mayo Clinic from 1951 through 1957 and compares these data with those from 219 patients with IE treated at Mayo Clinic 20 years later. These data are remarkably similar to those reported by other investigators.[276,277] The impact of improved microbiologic techniques is apparent in patients with Gram-negative bacillary, "other" infections, and "culture-negative" endocarditis. Some Gram-negative bacilli, such as *Haemophilus* species, are slow-growing microorganisms with special growth requirements. These fastidious microorganisms constitute a major portion of the Gram-negative bacillary cases of IE in the latter group of patients seen at Mayo Clinic. Increased sophistication of microbiologic techniques in the 1970s compared with the 1950s is at least partially accountable for the increased frequency of cases of Gram-negative bacillary IE seen during the 1970s. Of the 25 Mayo Clinic patients seen with IE in the 1970s caused by "other" microorganisms, 6 (24%) had infections with anaerobic Gram-positive cocci. In the 1950s anaerobic microbiology was largely a research tool, and patients with IE due to anaerobic bacteria would have been classified as "culture-negative."

Although improved techniques may explain some of the microbiologic differences between IE of the 1950s and the 1970s, the experience at Mayo Clinic and elsewhere indicates that the bacterial spectrum of IE is changing. The percentage of patients with viridans streptococcal IE declined from 53% in the 1950s to 37% in the 1970s; however, the average number of cases of IE increased from 24.5 per year 20 years ago to 44 per year during the past 5 years, and the number of patients with viridans streptococcal infections increased from 13 per year to 16 per year. It is clear that there are not fewer patients with viridans streptococcal IE, but, rather, that there are more patients with IE caused by other micro-

TABLE 3

Microbiology of Infective Endocarditis at Mayo Clinic

Organism	1951–1957 No. patients (%)[a]	1972–1976 No. patients (%)
Viridans streptococci	91 (53)	81 (37)
Group D streptococci	28 (16)	36 (16.5)
S. aureus	24 (14)	33 (15)
S. epidermidis	4 (2)	13 (6)
Gram-negative bacilli	10 (6)	24 (11)
Other microorganisms	1 (1)	25 (11.5)
Negative blood cultures	14 (8)	7 (3)
Total	172 (100)	219 (100)

[a]Geraci, J. E., The antibiotic therapy of bacterial endocarditis. Therapeutic data on 172 patients seen from 1951 through 1957: Additional observations on short-term therapy (two weeks) for penicillin-sensitive streptococcal endocarditis, *Med. Clin. North Am.*, July 1958, 1101.

From Wilson, W. R. and Washington, J. A., II, Infective endocarditis: a changing spectrum (editorial), *Mayo Clinc Proc.*, 54, 254, 1977. With permission.

organisms. Part of this changing spectrum is the result of nosocomially-acquired infections associated with the increased use of intravascular prostheses, life-support mechanisms and monitoring systems, and invasive diagnostic techniques. Another factor in the changing microbiologic spectrum of IE is the apparent increase in the number of patients with IE associated with intravenous drug abuse. Nosocomially acquired IE or IE associated with intravenous drug abuse is frequently due to Gram-negative bacilli, *Candida,* and "opportunistic" microorganisms, uncommon causes for IE in the 1950s.

Organisms Involved in Infective Endocarditis

Streptococci

Streptococci are the causative agent in 60 to 80% of cases of IE.[275-279] Viridans streptococci are numerically the predominant group of the streptococci and are the most common isolate from patients with IE. Group D streptococci are reported in some series to be the second leading cause of IE,[275] while other investigators state that group D streptococcal IE is third numerically as a cause of IE behind viridans streptococci and *S. aureus* infections.[276,277] Other streptococci, including anaerobic streptococci, group A β-hemolytic streptococci, group B streptococci, and other groups are isolated in the remainder of the cases of streptococcal IE.

Staphylococci

Infective endocarditis caused by *S. aureus* and *S. epidermidis* is second in frequency to that caused by streptococci. Staphylococci are responsible for 10 to 30% of the total number of cases of IE in most reported series. Nationwide, the incidence of *S. aureus* endocarditis has risen during the past 30 years.[276,280] Part of this increase is attributable to the high frequency of *S. aureus* endocarditis associated with intravenous drug abuse. A substantial number of cases of prosthetic valve endocarditis are caused by *S. epidermidis,*[186-188] which may partially explain the increasing frequency of IE caused by this microorganism.

Gram-negative Bacilli

Endocarditis caused by Gram-negative bacilli constitutes 5 to 11% of cases of IE.[276,277] At the Mayo Clinic from 1972 through 1976, *Haemophilus* species accounted for 37.5% of cases of Gram-negative bacillary IE.[275,281] Other investigators have reported *S. marcescens*[282] and *P. aeruginosa* IE[283] in association with intravenous drug abuse. Sporadic cases have been reported of Gram-negative bacillary endocarditis caused by *Brucella* species, *Campylobacter, E. coli, Klebsiella, Enterobacter, Alcaligenes faecalis, Cardiobacterium hominis, Eikenella corrodens, Pasteurella, Salmonella,* and other Gram-negative bacteria.[276,277] Gram-negative bacilli are also an important cause of prosthetic valve endocarditis.[186-188]

S. pneumoniae and N. gonorrhoeae

In the preantibiotic era, *S. pneumoniae* was responsible for 10% or more of cases of IE.[276,277] IE caused by this microorganism is now uncommon, accounting for <5% of cases. Right-sided pneumococcal endocarditis may occur in alcoholics in association with pneumonia and meningitis.[284] Like *S. pneumoniae, N. gonorrhoeae* was once a common cause of infective endocarditis (5 to 10% of cases),[276,277,285] but with the advent of effective antimicrobial therapy for gonorrhea, gonococcal infective endocarditis is now relatively rare.

Fungi

Candida and *Aspergillus* species are the most common causes of fungal endocarditis.[277,286] *Histoplasma, Blastomyces, Cryptococcus,* and *Torulopsis* have also been reported to cause infective endocarditis.[277] Fungal endocarditis occurring in association with intravenous drug abuse is most often caused by species other than *Candida albicans* and most often involves the left side of the heart,[287] although *Candida* and *Aspergillus* species may cause prosthetic valve endocarditis.[186-188] Fungal vegetations are usually large and friable, and systemic embolization, especially to the lower extremities, is common. Blood cultures are usually negative in patients with *Aspergillus* endocarditis and are frequently positive in patients with *Candida* endocarditis.

Other Microorganisms

A multiplicity of other microorganisms has been reported to cause infective endocarditis, including *B. fragilis* and other anaerobes, *Corynebacterium, Listeria monocytogenes, Erysipelothrix, Spirillum minus, Coxiella burnetii,* and other organisms.[276,277]

Prosthetic Valve Endocarditis

Prosthetic valve endocarditis (PVE) is an infrequent, serious complication of cardiac valve replacement,[186-188] reported to occur in 1 to 3% of patients undergoing cardiac valve replacement. The microbiologic spectrum of PVE differs from that of natural valve endocarditis and should be considered separately. Patients with PVE are divided into two groups: those with early occurrence of PVE (within 2 months postoperatively) and those with late occurrence (2 or more months postoperatively). The microbiologic etiology of 45 patients with PVE seen at Mayo Clinic during an 11-year period from 1963 to 1973 is listed in Table 4.[188] *S. aureus* was the most frequent isolate in patients with early-onset infections, viridans streptococci the leading cause of late-onset infections, and Gram-negative bacilli the second leading cause of PVE in both early- and late-onset infections. *S. epidermidis* and other "opportunistic" bacteria, such as corynebacteria, have been isolated frequently from patients with early- and late-onset PVE.

The role in PVE of possible intraoperative contamination of prosthetic valves with "opportunistic" bacteria has been emphasized by Blakemore et al.[189] In their study of 32 patients undergoing cardiopulmonary bypass, *S. epidermidis*, corynebacteria, streptococci, Gram-negative bacilli, and fungi were isolated from the cardiopulmonary bypass apparatus in 75% of the cases. Ankeney and Parker[190] isolated *S. epidermidis* from 7.5% of 1555 blood cultures obtained from 5 different sites during 383 open-heart procedures; the 2 sites yielding the most positive cultures were the extracorporeal pump tubing and the coronary sucker.

TABLE 4

Infecting Microorganisms in Patients with Prosthetic Valve Endocarditis

	Isolations			
	Early group		Late group	
Organism	No.	%	No.	%
Staphylococcus aureus	7	44	3	10
Gram-negative bacilli	6	38	9	31
Streptococci, viridans group	0	0	12[a]	41
Streptococci, Group D	1	6	3	10
Staphylococcus epidermidis	1	6	1	3
Corynebacterium xerosis	1	6	1	3
Candida parapsilosis	0	0	1	3

[a]Including a second organism (*Alcaligenes faecalis*) in one patient.

Modified from Wilson, W. R., Jaumin, P. M., Danielson, G. K., Giuliani, E. R., Washington, J. A., II, and Geraci, J. E., *Ann. Intern. Med.,* 82, 751, 1975. With permission.

Culture-negative Endocarditis

Culture-negative IE has been reported to occur in 3 to 14% of cases.[276–278] In recent years, the frequency of cases of culture-negative IE has declined,[276] with improved microbiologic techniques at least partially responsible. In the 1950s, anaerobic microbiology was largely a research tool, and patients with IE due to anaerobic bacteria would have been classified as culture negative. Some of the viridans streptococci and Gram-negative bacilli, such as *Haemophilus* species, are fastidious slow-growing microorganisms which require special growth factors or increased carbon dioxide, and, formerly, IE caused by these microorganisms might have been considered culture negative. Currently, partially treated cases of IE are probably the most common causes of culture-negative endocarditis.

SIGNIFICANCE OF FINDINGS

The clinical significance of a positive blood culture ranges from a trivial, asymptomatic, transient bacteremia related to a manipulative procedure to profound sepsis, shock, and death. In the former group of patients, difficulty may arise in differentiating between a transient bacteremia and a blood culture contaminant. In the study by Hall et al.,[4] it should be noted that 98 of the 116 presumed contaminants of blood cultures were aerobic or anaerobic corynebacteria. In this study, blood cultures were obtained from patients suspected of having sepsis. These microorganisms were not isolated from blood cultures obtained from normal, healthy adult outpatients.[3] Although these two populations of patients are not comparable, the methods used to collect and process the blood cultures were the same. Therefore, the absence of isolation of corynebacteria from the normal control population might lead one to question the assumption that these organisms usually are contaminants. When *Corynebacterium* or *S. epidermidis* is isolated from patients with suspected IE and patients with prosthetic heart valves, central nervous system shunts, or other prosthetic devices, especially if isolated from multiple blood cultures, they should be considered clinically significant.

While the significance of the isolation from blood cultures of organisms which are part of the normal skin microflora is often not clear, the isolation of streptococci in two or more blood cultures, or *S. aureus* in one or more blood cultures, is usually indicative of clinically significant infection. Patients with repeated isolations of viridans streptococci or nonenterococcal group D streptococci from blood cultures should be considered to have IE until proven otherwise. Clinicians should also suspect IE in patients with enterococcal or *S. aureus* bacteremia when no portal of entry is clearly identifiable. The isolations from blood cultures of *S. pneumoniae,* group A β-hemolytic streptococci, and group B streptococci are almost always clinically significant and may be associated with respiratory infection, puerperal sepsis, infective endocarditis, meningitis, etc.

Patients with blood cultures positive for Gram-negative bacilli should be considered to have clinically significant infections. The most important factors in the prognosis of patients with Gram-negative bacillary septicemia are the underlying disease of the host and the precipitating factor(s) resulting in the septicemia. McCabe[146] classified patients with Gram-negative bacteremias into three groups: (1) patients with rapidly fatal underlying disease which could be expected to cause death within 1 year (i.e., patients with acute leukemia), (2) patients with ultimately fatal underlying disease with which survival exceeding 4 years would not be anticipated (i.e., patients with metastatic malignancy, lymphoma, cirrhosis with bleeding, esophageal varices, etc.), and (3) patients with nonfatal underlying disease which was not anticipated to cause death during the next 4 years. In this study of 173 patients with Gram-negative bacteremia, the severity of the underlying disease was a more important determinant of survival than the species of bacteria involved or the type of antimicrobial therapy administered. These observations have been confirmed by other investigators.[140,145,239] Other important prognostic factors are the patient's age, whether the infection is community- or hospital-acquired, and whether septic shock and azotemia develop. Polymicrobial bacteremia is associated with a higher mortality rate than unimicrobial bacteremia. Hermans and Washington[172] reported 37% mortality among 46 patients with polymicrobial bacteremia compared with 20% mortality in patients with unimicrobial Gram-negative bacteremia ($P < 0.05$). In this study, the highest mortality occurred in patients with severe underlying disease and polymicrobial bacteremia.

The factors important in the prognosis of

patients with septicemia due to Gram-negative bacilli are also important in the prognosis of patients with sepsis caused by Gram-positive cocci and other microorganisms. The overall mortality in patients with *S. aureus* septicemia is reported to be 21 to 42%.[288,289] Patients with *S. aureus* septicemia have a favorable prognosis when there is a readily identifiable focus of infection which may be eradicated (i.e., infected intravenous catheter, subcutaneous abscess, etc.). Contrariwise, patients with *S. aureus* PVE have a high mortality – 90% in one study.[188]

Since the advent of antimicrobial therapy, the mortality associated with *S. pneumoniae* sepsis has decreased considerably.[140] However, it remains high in patients with severe underlying disease. A particularly fulminant, usually fatal form of *S. pneumoniae* septicemia occurs in patients who have undergone splenectomy prior to renal transplantation or for staging of lymphoproliferative disorders.[255,256]

The mortality of patients with *B. fragilis* septicemia (44%) is higher than that of patients with aerobic Gram-negative bacillary sepsis (20%).[172,173] One reason for this may be the frequent association between *Bacteroides* septicemia and carcinoma of the colon and other gastrointestinal disorders. Another reason for the higher mortality may be that polymicrobial bacteremia occurs frequently in patients with *Bacteroides* septicemia – 31% of patients in one study – whereas only 6% of bacteremic patients reported by Hermans and Washington had polymicrobial bacteremia.[173]

Most patients with blood cultures positive for *B. fragilis* had clinically significant bacteremia.[173] The clinical spectrum of anaerobic bacteremia caused by other microorganisms ranges from asymptomatic bacteremia to profound sepsis with hemolytic anemia, jaundice, renal failure, septic shock, and death.[174] This broad spectrum is particularly apparent in patients with clostridial bacteremia. Several studies have emphasized the frequently benign clinical course of clostridial bacteremia.[246,274] Septic shock syndrome associated with clostridial bacteremia occurs most frequently in patients with septic abortion or patients with devitalized tissue and gas gangrene.[174] Occasionally, the isolation of *Candida* species from blood cultures may represent contamination; however, *Candida* from repeated blood cultures, or from immunosuppressed patients, is usually indicative of a clinically significant fungemia.[290]

Not all patients with candidemia, however, require antifungal therapy. In nonimmunosuppressed patients, the use of broad-spectrum antimicrobials and intravenous hyperalimentation predisposes patients to candidemia. Simply removing an infected intravenous catheter or discontinuing broad-spectrum antimicrobials may be sufficient therapy for fungemia. Fungal blood cultures should be repeated at intervals thereafter to insure that fungemia has ceased. *Candida* ophthalmitis may occur, and funduscopic examination should be performed regularly for at least 1 month following a *Candida* fungemia.[291] In immunosuppressed patients, those with multiple blood cultures positive for *Candida* after infected intravenous catheters have been removed or after broad-spectrum antimicrobials have been discontinued, or with *Candida* ophthalmitis, antifungal therapy with amphotericin B and possibly 5-fluorocytosine may be indicated. While the significance of a blood culture positive for *Candida* is not always clear, the isolation from blood cultures of other fungi, such as *Histoplasma capsulatum, Cryptococcus neoformans,* and *Torulopsis glabrata,* should always be considered clinically significant. Patients with *Nocardia* bacteremia should be suspected of having metastatic abscesses, especially in the brain.[265]

In summary, each patient with a positive blood culture must be considered individually, and the significance of a positive blood culture must be assessed by considering the composite clinical condition of the patient. It is important that clinicians not overreact to the report of the positive blood culture which may represent contamination or transient, self-limited infection. Such overreaction may result in unnecessary treatment with attendant increased cost to the patient and iatrogenic morbidity and, possibly, mortality. Equally important, the clinician should not underreact. Infections are among the leading causes of death occurring in hospitalized patients, and patients with clinically significant bacteremia require aggressive therapy with antimicrobials plus appropriate adjunctive measures. Physicians must, therefore, acquire a clear understanding of predisposing conditions and underlying diseases of patients with septicemia and familiarize themselves with the microbiologic and other laboratory tools essential to the management of these patients.

REFERENCES

1. **Reith, A. F. and Squier, T. L.,** Blood cultures of apparently healthy persons, *J. Infect. Dis.,* 51, 336, 1932.
2. **MacGregor, R. R. and Beaty, H. N.,** Evaluation of positive blood cultures: guidelines for early differentiation of contaminated from valid positive cultures, *Arch. Intern. Med.,* 130, 84, 1972.
3. **Wilson, W. R., Van Scoy, R. E., and Washington, J. A., II,** Incidence of bacteremia in adults without infection, *J. Clin. Microbiol.,* 2, 94, 1975.
4. **Hall, M., Warren, E., and Washington, J. A., II,** Comparison of two liquid blood culture media containing sodium polyanetholesulfonate: tryptic soy and Columbia, *Appl. Microbiol.,* 27, 699, 1974.
5. **Elliott, S. D.,** Bacteriaemia and oral sepsis, *Proc. R. Soc. Med.,* 32, 747, 1939.
6. **Schirger, A., Martin, W. J., Royer, R. Q., and Needham, G. M.,** Bacterial invasion of blood after oral surgical procedures, *J. Lab. Clin. Med.,* 55, 376, 1960.
7. **Scott, W. W.,** Blood stream infections in urology: a report of eighty-two cases, *J. Urol.,* 21, 527, 1929.
8. **Barrington, F. J. F. and Wright, H. D.,** Bacteriaemia following operations on the urethra, *J. Pathol. Bacteriol.,* 33, 871, 1930.
9. **Powers, J. H.,** Bacteriemia following instrumentation of the infected urinary tract, *N.Y. State J. Med.,* 36, 323, 1936.
10. **Biorn, C. L., Browning, W. H., and Thompson, L.,** Transient bacteremia immediately following transurethral prostatic resection, *J. Urol.,* 63, 155, 1950.
11. **Bulkley, G. J., O'Conor, V. J., and Sokol, J. K.,** A clinical study of bacteremia and overhydration following transurethral resection, *J. Urol.,* 72, 1205, 1954.
12. **Creevy, C. D. and Feeney, M. J.,** Routine use of antibiotics in transurethral prostatic resection: a clinical investigation, *J. Urol.,* 71, 615, 1954.
13. **Appleton, D. M. and Waisbren, B. A.** The prophylactic use of chloramphenicol in transurethral resections of the prostate gland, *J. Urol.,* 75, 304, 1956.
14. **Slade, N.,** Bacteriaemia and septicaemia after urological operations, *Proc. R. Soc. Med.,* 51, 331, 1958.
15. **Kidd, E. E. and Burnside, K.,** Bacteraemia, septicaemia and intravascular haemolysis during transurethral resection of the prostate gland, *Br. J. Urol.,* 37, 551, 1965.
16. **Last, P. M., Harbison, P. A., and Marsh, J. A.,** Bacteraemia after urological instrumentation, *Lancet,* 1, 74, 1966.
17. **Sullivan, N. M., Sutter, V. L., Carter, W. T., Attebery, H. R., and Finegold, S. M.,** Bacteremia after genitourinary tract manipulation: bacteriological aspects and evaluation of various blood culture systems, *Appl. Microbiol.,* 23, 1101, 1972.
18. **Sullivan, N. M., Sutter, V. L., Mims, M. M., Marsh, V. H., and Finegold, S. M.,** Clinical aspects of bacteremia after manipulation of the genitourinary tract, *J. Infect. Dis.,* 127, 49, 1973.
19. **LeFrock, J. L., Ellis, C. A., Turchik, J. B., and Weinstein, L.,** Transient bacteremia associated with sigmoidoscopy, *N. Engl. J. Med.,* 289, 467, 1973.
20. **Liebermann, T. R.,** Bacteremia and fiberoptic endoscopy, *Gastrointest. Endosc.,* 23, 36, 1976.
21. **Pelican, G., Hentges, D., Butt, J., Haag, T., Rolfe, R., and Hutcheson, D.,** Bacteremia during colonoscopy, *Gastrointest. Endosc.,* 23, 33, 1976.
22. **Rafoth, R. J., Sorenson, R. M., and Bond, J. H., Jr.,** Bacteremia following colonoscopy, *Gastrointest. Endosc.,* 22, 32, 1975.
23. **McCloskey, R. V., Gold, M., and Weser, E.,** Bacteremia after liver biopsy, *Arch. Intern. Med.,* 132, 213, 1973.
24. **Le Frock, J. L., Ellis, C. A., Turchik, J. B., Zawacki, J. K., and Weinstein, L.,** Transient bacteremia associated with percutaneous liver biopsy, *J. Infect. Dis.,* Suppl. 131, 104, 1975.
25. **LeFrock, J. L., Klainer, A. S., Wu, W. -H., and Turndorf, H.,** Transient bacteremia associated with nasotracheal suctioning, *JAMA,* 236, 1610, 1976.
26. **Raines, D. R., Branche, W. C., Anderson, D. L., and Boyce, H. W., Jr.,** The occurrence of bacteremia after esophageal dilation, *Gastrointest. Endosc.,* 22, 86, 1975.
27. **Le Frock, J., Ellis, C. A., Klainer, A. S., and Weinstein, L.,** Transient bacteremia associated with barium enema, *Arch. Intern. Med.,* 135, 835, 1975.
28. **Timms, R. M. and Harrell, J. H.,** Bacteremia related to fiberoptic bronchoscopy: a case report, *Am. Rev. Respir. Dis.,* 111, 555, 1975.
29. **Shawker, T. H., Kluge, R. M., and Ayella, R. J.,** Bacteremia associated with angiography, *JAMA,* 229, 1090, 1974.
30. **Sande, M. A., Levinson, M. E., Lukas, D. S., and Kaye, D.,** Bacteremia associated with cardiac catheterization, *N. Engl. J. Med.,* 281, 1104, 1969.
31. **Siroky, M. B., Moylan, R. A., Austen, G., Jr., and Olsson, C. A.,** Metastatic infection secondary to genitourinary tract sepsis, *Am. J. Med.,* 61, 351, 1976.
32. **Mendelssohn, K. and Witts, L. J.,** Transmission of infection during withdrawal of blood, *Br. Med. J.,* 1, 625, 1945.
33. **Katz, L., Johnson, D. L., Neufeld, P. D., and Gupta, K. G.,** Evacuated blood-collection tubes: the backflow hazard, *Can. Med. Assoc. J.,* 113, 208, 1975.
34. **McLeish, W. A., Corrigan, E. N., Elder, R. H., and Westwood, J. C. N.,** Contaminated vacuum tubes (letter to the editor), *Can. Med. Assoc. J.,* 112, 682, 1975.
35. **Washington, J. A., II,** The microbiology of evacuated blood collection tubes, *Ann. Intern. Med.,* 86, 186, 1977.

36. Maki, D. G., Goldman, D. A., and Rhame, F. S., Infection control in intravenous therapy, *Ann. Intern. Med.*, 79, 867, 1973.
37. Collins, R. N., Braun, P. A., Zinner, S. H., and Kass, E. H., Risk of local and systemic infection with polyethylene intravenous catheters: a prospective study of 213 catherizations, *N. Engl. J. Med.*, 279, 340, 1968.
38. Bentley, D. W. and Lepper, M. H., Septicemia related to indwelling venous catheter, *JAMA*, 206, 1749, 1968.
39. Zinner, S. H., Denny-Brown B. C., Braun, P., Burke, J. P., Toala, P., and Kass, E. H., Risk of infection with intravenous indwelling catheters: effect of application of antibiotic ointment, *J. Infect. Dis.*, 120, 616, 1969.
40. Norden, C. W., Application of antibiotic ointment to the site of venous catheterization: a controlled trial, *J. Infect. Dis.*, 120, 611, 1969.
41. Banks, D. C., Yates, D. B., Cawdrey, H. M., Harries, M. G., and Kidner, P. H., Infection from intravenous catheters, *Lancet*, 1, 443, 1970.
42. Bolasny, B. L., Shepard, G. H., and Scott, H. W., Jr., The hazards of intravenous polyethylene catheters in surgical patients, *Surg. Gynecol. Obstet.*, 130, 342, 1970.
43. Levy, R. S., Goldstein, J., and Pressman, R. S., Value of a topical antibiotic ointment in reducing bacterial colonization of percutaneous venous catheters, *J. Albert Einstein Med. Cent.*, 18, 67, 1970.
44. Bernard, R. W., Stahl, W. M., and Chase, R. M., Jr., Subclavian vein catheterizations: a prospective study. II. Infectious complications, *Ann. Surg.*, 173, 191, 1971.
45. Hoshal, V. L., Jr., Intravenous catheters and infection, *Surg. Clin. North Am.*, 52, 1407, 1972.
46. Mogensen, J. V., Frederiksen, W., and Jensen, J. K., Subclavian vein catheterization and infection: a bacteriological study of 130 catheter insertions, *Scand. J. Infect. Dis.*, 4, 31, 1972.
47. Peter, G., Lloyd-Still, J. D., and Lovejoy, F. H., Jr., Local infection and bacteremia from scalp vein needles and polyethylene catheters in children, *J. Pediatr.*, 80, 78, 1972.
48. Moran, J. M., Atwood, R. P., and Rowe, M. I., A clinical and bacteriologic study of infections associated with venous cutdowns, *N. Engl. J. Med.*, 272, 554, 1965.
49. Druskin, M. S. and Siegel, P. D., Bacterial contamination of indwelling intravenous polyethylene catheters, *JAMA*, 185, 966, 1963.
50. Cheney, F. W., Jr. and Lincoln, J. R., Phlebitis from plastic intravenous catheters, *Anesthesiology*, 25, 650, 1964.
51. Brereton, R. B., Incidence of complications from indwelling venous catheters, *Del. Med. J.*, 41, 1, 1969.
52. Corso, J. A., Agostinelli, R., and Brandriss, M. W., Maintenance of venous polyethylene catheters to reduce risk of infection, *JAMA*, 210, 2075, 1969.
53. Glover, J. L., O'Byrne, S. A., and Jolly, L., Infusion catheter sepsis: an increasing threat, *Ann. Surg.*, 173, 148, 1971.
54. Fuchs, P. C., Indwelling intravenous polyethylene catheters: factors influencing the risk of microbial colonization and sepsis, *JAMA*, 216, 1447, 1971.
55. Bolasny, B. L., Martin, C. E., and Conkle, D. M., Careful technique with plastic intravenous catheters, *Surg. Gynecol. Obstet.*, 132, 1030, 1971.
56. Crenshaw, C. A., Kelly, L., Turner, R. J., III, and Enas, D., Bacteriologic nature and prevention of contamination to intravenous catheters, *Am. J. Surg.*, 123, 264, 1972.
57. Freeman, R. and King, B., Infective complications of indwelling intravenous catheters and the monitoring of infections by the nitroblue-tetrazolium test, *Lancet*, 1, 992, 1972.
58. Keye, J. D., Jr. and Magee, W. E., Fungal diseases in a general hospital: a study of 88 patients, *Am. J. Clin. Pathol.*, 26, 1235, 1956.
59. Dobias, B., Moniliasis in pediatrics, *Am. J. Dis. Child.*, 94, 234, 1957.
60. Vince, S., The spread of *Candida* in infants and children, *Med. J. Aust.*, 2, 143, 1959.
61. Louria, D. B., Stiff, D. P., and Bennett, B., Disseminated moniliasis in the adult, *Medicine*, 41, 307, 1962.
62. Bodey, G. P., Fungal infections complicating acute leukemia, *J. Chronic Dis.*, 19, 667, 1966.
63. White, J. J., Wallace, C. K., and Burnett, L. S., Drug letter: skin disinfection, *Johns Hopkins Med. J.*, 126, 169, 1970.
64. Gershenfeld, L., Iodine, in *Disinfection, Sterilization, and Preservation,* Lawrence, C. A. and Block, S. S., Eds., Lea & Febiger, Philadelphia, 1968, 329.
65. Lowbury, E. J. L., Lilly, H. A., and Bull, J. P., Disinfection of the skin of operation sites, *Br. Med. J.*, 2, 1039, 1960.
66. Price, P. B., Surgical antiseptics, in *Disinfection, Sterilization, and Preservation,* Lawrence, C. A. and Block, S. S., Eds., Lea & Febiger, Philadelphia, 1968, 532.
67. Selwyn, S. and Ellis, H., Skin bacteria and skin disinfection reconsidered, *Br. Med. J.*, 1, 136, 1972.
68. Joress, S. M., A study of disinfection of the skin: a comparison of Povidone-iodine with other agents used for surgical scrubs, *Ann. Surg.*, 155, 296, 1962.
69. O'Neill, J. A., Jr., Pruitt, B. A., Jr., Foley, F. D., and Moncrief, J. A., Suppurative thrombophlebitis: a lethal complication of intravenous therapy, *J. Trauma*, 8, 256, 1968.
70. Stein, J. M. and Pruitt, B. A., Jr., Suppurative thrombophlebitis: a lethal iatrogenic disease, *N. Engl. J. Med.*, 282, 1452, 1970.
71. Pruitt, B. A., Jr., Stein, J. M., Foley, F. D., Moncrief, J. A., and O'Neill, J. A., Jr., Intravenous therapy in burn patients: suppurative thrombophlebitis and other life-threatening complications, *Arch. Surg.*, 100, 399, 1970.

72. Editorial, Septic thrombophlebitis and venous cannulas, *Lancet,* 2, 406, 1970.
73. **Foley, F. D.,** The burn autopsy: fatal complications of burns, *Am. J. Clin. Pathol.,* 52, 1, 1969.
74. **Moncrief, J. A.,** Femoral catheters, *Ann. Surg.,* 147, 166, 1958.
75. **Bansmer, G., Keith, D., and Tesluk, H.,** Complications following use of indwelling catheters of inferior vena cava, *JAMA,* 167, 1606, 1958.
76. **Indar, R.,** The dangers of indwelling polyethylene cannulae in deep veins, *Lancet,* 1, 284, 1959.
77. **Crane, C.,** Venous interruption for septic thrombophlebitis, *N. Engl. J. Med.,* 262, 947, 1960.
78. **McNair, T. J. and Dudley, H. A. F.,** The local complications of intravenous therapy, *Lancet,* 2, 365, 1959.
79. **Collins, H. S., Helper, A. N., Blevins, A., and Olenberg, G.,** Staphylococcal bacteremia, *Ann. N.Y. Acad. Sci.,* 65, 222, 1956.
80. **Maki, D. G., Rhame, F. S., Mackel, D. C., and Bennett, J. V.,** Nationwide epidemic of septicemia caused by contaminated intravenous products. I. Epidemiologic and clinical features, *Am. J. Med.,* 60, 471, 1976.
81. Center for Disease Control, Nosocomial bacteremias associated with intravenous fluid therapy – USA, *Morbid. Mortal. Weekly Rep.,* No. 20, Suppl. 9, 81, 1971.
82. Center for Disease Control, Follow-up on septicemias associated with contaminated Abbott intravenous fluids: United States, *Morbid. Mortal. Weekly Rep.,* 20, 91, 1971.
83. Center for Disease Control, Follow-up on septicemia associated with contaminated intravenous fluid from Abbott Laboratories, *Morbid. Mortal. Weekly Rep.,* 20, 110, 1971.
84. **Maki, D. G., Rhame, R. S., Goldmann, D. A., and Mandell, G. L.,** The infection hazard posed by contaminated intravenous infusion fluid, in *Bacteremia: Laboratory and Clinical Aspects,* Sonnenwirth, A. C., Ed., Charles C Thomas, Springfield, Ill., 1973, 76.
85. **Mackel D. C., Maki, D. G., Anderson, R. A., Rhame, F. S., and Bennett, J. V.,** Mechanisms of contamination of a screw-cap closure for infusion fluid, Presented at 11th Interscience Conf. on Antimicrobial Agents and Chemotherapy, Atlantic City, October 20, 1971.
86. **Felts, S. K., Schaffner, W., Melly, M. A., and Koenig, M. G.,** Sepsis caused by contaminated intravenous fluids: epidemiologic, clinical, and laboratory investigation of an outbreak in one hospital, *Ann. Intern. Med.,* 77, 881, 1972.
87. **Michaels, L. and Ruebner, B.,** Growth of bacteria in intravenous infusion fluids, *Lancet,* 1, 772, 1953.
88. **Sack, R. A.,** Epidemic of gram-negative organism septicemia subsequent to elective operation, *Am. J. Obstet. Gynecol.,* 107, 394, 1970.
89. **Duma, R. J., Warner, J. F., and Dalton, H. P.,** Septicemia from intravenous infusions, *N. Engl. J. Med.,* 284, 257, 1971.
90. **Dudrick, S. J., Groff, D. B., and Wilmore, D. W.,** Long term venous catheterization in infants, *Surg. Gynecol. Obstet.,* 129, 805, 1969.
91. **Wilmore, D. W. and Dudrick, S. J.,** Safe long-term venous catheterization, *Arch. Surg.,* 98, 256, 1969.
92. **Groff, D. B.,** Complications of intravenous hyperalimentation in newborns and infants, *J. Pediatr. Surg.,* 4, 460, 1969.
93. **McGovern, B.,** Intravenous hyperalimentation, *Mil. Med.,* 135, 1137, 1970.
94. **Ashcraft, K. W. and Leape, L. L.,** *Candida* sepsis complicating parenteral feeding, *JAMA,* 212, 454, 1970.
95. **Curry, C. R. and Quie, P. G.,** Fungal septicemia in patients receiving parenteral hyperalimentation, *N. Engl. J. Med.,* 285, 1221, 1971.
96. **Peden, V. H. and Karpel, J. T.,** Total parenteral nutrition in premature infants, *J. Pediatr.,* 81, 137, 1972.
97. **Winters, R. W., Santulli, T. V., Heird, W. C., Schullinger, J. N., and Driscoll, J. M., Jr.,** Hyperalimentation without sepsis (letter to the editor), *N. Engl. J. Med.,* 286, 321, 1972.
98. **Goldmann, D. A. and Maki, D. G.,** Infection control in total parenteral nutrition, *JAMA,* 223, 1360, 1973.
99. **Owings, J. M., Bomar, W. E., Jr., and Ramage, R. C.,** Parenteral hyperalimentation and its practical applications, *Ann. Surg.,* 175, 712, 1972.
100. **Filler, R. M.,** Total parenteral feeding of infants, *Hosp. Pract.,* 7, 79, 1972.
101. **Parsa, M. H., Habif, D. V., Ferrer, J. M., Lipton, R., and Yoshimura, N. N.,** Intravenous hyperalimentation: indications, technique, and complications, *Bull. N.Y. Acad. Med.,* 48, 920, 1972.
102. **Freeman, J. B., Lemire, A., and MacLean, L. D.,** Intravenous alimentation and septicemia, *Surg. Gynecol. Obstet.,* 135, 708, 1972.
103. **Dillon, J. D., Jr., Schaffner, W., Van Way, C. W., III, and Meng, H. C.,** Septicemia and total parenteral nutrition: distinguishing catheter-related from other septic episodes, *JAMA,* 223, 1341, 1973.
104. **Sanderson, I. and Deitel, M.,** Intravenous hyperalimentation without sepsis, *Surg. Gynecol. Obstet.,* 136, 577, 1973.
105. **Sanders, R. A. and Sheldon, G. F.,** Septic complications of total parenteral nutrition: a five year experience, *Am. J. Surg.,* 132, 214, 1976.
106. **Fischer, J. E., Abbott, W. M., and Abel, R. M.,** Fungal septicaemia complicating intravenous hyperalimentation (letter to the editor), *Lancet,* 1, 640, 1972.
107. **Neuhof, H. and Seley, G. P.,** Acute suppurative phlebitis complicated by septicemia, *Surgery,* 21, 831, 1947.
108. **Phillips, R. W. and Eyre, J. D., Jr.,** Septic thrombophlebitis with septicemia: report of three cases due to *Staphylococcus aureus* infection after the intravenous use of polyethylene catheters for parenteral therapy, *N. Engl. J. Med.,* 259, 729, 1958.

109. **Fisher, E. J., Maki, D. G., Eisses, J., Neblett, T. R., and Quinn, E. L.,** Epidemic septicemias due to intrinsically contaminated infusion products, presented at the 11th Interscience Conf. on Antimicrobial Agents and Chemotherapy, Atlantic City, October 20, 1971.
110. Center for Disease Control, Septicemias associated with contaminated intravenous fluids – Wisconsin, Ohio, *Morbid. Mortal. Weekly Rep.*, 22, 99, 1973.
111. Center for Disease Control, Follow-up on septicemias associated with contaminated intravenous fluids – United States, *Morbid. Mortal. Weekly Rep.*, 22, 115, 1973.
112. Center for Disease Control, Follow-up on septicemias associated with contamination of intravenous fluids – United States, *Morbid. Mortal. Weekly Rep.*, 22, 124, 1973.
113. **Phillips, I., Eykyn, S., and Laker, M.,** Outbreak of hospital infection caused by contaminated autoclaved fluids, *Lancet*, 1, 1258, 1972.
114. **Smits, H. and Freedman, L. R.,** Prolonged venous catheterization as a cause of sepsis, *N. Engl. J. Med.*, 276, 1229, 1967.
115. **Darrell, J. H. and Garrod, L. P.,** Secondary septicaemia from intravenous cannulae, *Br. Med. J.*, 2, 481, 1969.
116. **Henzel, J. H. and DeWeese, M. S.,** Morbid and mortal complications associated with prolonged central venous cannulation: awareness, recognition, and prevention, *Am. J. Surg.*, 121, 600, 1971.
117. **Walters, M. B., Stanger, H. A. D., and Rotem, C. E.,** Complications with percutaneous central venous catheters, *JAMA*, 220, 1455, 1972.
118. **Mays, E. T.,** A microbiologic investigation of percutaneous central venous catheters, *South. Med. J.*, 65, 830, 1972.
119. **Colvin, M. P., Blogg, C. E., Savege, T. M., Jarvis, J. D., and Strunin, L.,** A safe long-term infusion technique? *Lancet*, 2, 317, 1972.
120. **Sketch, M. H., Cale, M., Mohiuddin, S. M., and Booth, R. W.,** Use of percutaneously inserted venous catheters in coronary care units, *Chest*, 62, 684, 1972.
121. **Buchsbaum, H. J. and White, A. J.,** The use of subclavian central venous catheters in gynecology and obstetrics, *Surg. Gynecol. Obstet.*, 136, 561, 1973.
122. **Krauss, A. N., Albert, R. F., and Kannan, M. M.,** Contamination of umbilical catheters in the newborn infant, *J. Pediatr.*, 77, 965, 1970.
123. **Balagtas, R. C., Bell, C. E., Edwards, L. D., and Levin, S.,** Risk of local and systemic infections associated with umbilical vein catheterization: a prospective study in 86 newborn patients, *Pediatrics*, 48, 359, 1971.
124. **Krauss, A. N., Caliendo, T. J., and Kannan, M. M.,** Bacteremia in newborn infants: following umbilical catheterization, *N.Y. State J. Med.*, 72, 1136, 1972.
125. **Casalino, M. B. and Lipsitz, P. J.,** Contamination of umbilical catheters (letter to the editor), *J. Pediatr.*, 78, 1077, 1971.
126. **Powers, W. F. and Tooley, W. H.,** Contamination of umbilical vessel catheters: encouraging information (letter to the editor), *Pediatrics*, 49, 470, 1972.
127. **Symansky, M. R. and Fox, H. A.,** Umbilical vessel catheterization: indications, management, and evaluation of the technique, *J. Pediatr.*, 80, 820, 1972.
128. **Lowenbraun, S., Young, V., Kenton, D., and Serpick, A. A.,** Infection from intravenous "scalp-vein" needles in a susceptible population, *JAMA*, 212, 451, 1970.
129. **Lloyd-Still, J. D., Peter, G., and Lovejoy, F. H., Jr.,** Infected "scalp-vein" needles (letter to the editor), *JAMA*, 213, 1496, 1970.
130. **Crossley, K. and Matsen, J. M.,** The scalp-vein needle: a prospective study of complications, *JAMA*, 220, 985, 1972.
131. **Borden, C. W. and Hall, W. H.,** Fatal transfusion reactions from massive bacterial contamination of blood, *N. Engl. J. Med.*, 245, 760, 1951.
132. **Stevens, A. R., Jr., Legg, J. S., Henry, B. S., Dille, J. M., Kirby, W. M. M., and Finch, C. A.,** Fatal transfusion reactions from contamination of stored blood by cold growing bacteria, *Ann. Intern. Med.*, 39, 1228, 1953.
133. **McEntegart, M. G.,** Dangerous contaminants in stored blood, *Lancet*, 2, 909, 1956.
134. **Braude, A. I.,** Transfusion reactions from contaminated blood: their recognition and treatment, *N. Engl. J. Med.*, 258, 1289, 1958.
135. **Habibi, B. and Salmon, C.,** Septic shock from bacterial contamination of transfused blood (letter to the editor), *Lancet*, 2, 830, 1972.
136. **Braude, A. I., Sanford, J. P., Bartlett, J. E., and Mallery, O. T., Jr.,** Effects and clinical significance of bacterial contaminants in transfused blood, *J. Lab. Clin. Med.*, 39, 902, 1952.
137. **James, J. D.,** Bacterial contamination of preserved blood (editorial), *Vox. Sang.*, 4, 177, 1959.
138. **Buchholz, D. H., Young, V. M., Friedman, N. R., Reilly, J. A., and Mardiney, M. R., Jr.,** Bacterial proliferation in platelet products stored at room temperature: transfusion-induced enterobacter sepsis, *N. Engl. J. Med.*, 285, 429, 1971.
139. **Braude, A. I., Carey, F. J., and Siemienski, J.,** Studies of bacterial transfusion reactions from refrigerated blood: the properties of cold-growing bacteria, *J. Clin. Invest.*, 34, 311, 1955.
140. **McGowan, J. E., Jr., Barnes, M. W., and Finland, M.,** Bacteremia at Boston City Hospital: occurrence and mortality during 12 selected years (1935–1972), with special reference to hospital-acquired cases, *J. Infect. Dis.*, 132, 316, 1975.

141. **Tillotson, J. R. and Finland, M.,** Bacterial colonization and clinical superinfection of the respiratory tract complicating antibiotic treatment of pneumonia, *J. Infect. Dis.,* 119, 597, 1969.
142. **Flick, M. R. and Cluff, L. E.,** Pseudomonas bacteremia: review of 108 cases, *Am. J. Med.,* 60, 501, 1976.
143. **McHenry, M. C., Alfidi, R. J., Deodhar, S. D., Braun, W. E., and Popowniak, K. L.,** Hospital-acquired pneumonia, *Med. Clin. North Am.,* 58, 565, 1974.
144. **Pierce, A. K., Sanford, J. P., Thomas, G. D., and Leonard, J. S.,** Long-term evaluation of decontamination of inhalation-therapy equipment and the occurrence of necrotizing pneumonia, *N. Engl. J. Med.,* 282, 528, 1970.
145. **McHenry, M. C. and Hawk, W. A.,** Bacteremia caused by gram-negative bacilli, *Med. Clin. North Am.,* 58, 623, 1974.
146. **McCabe, W. R.,** Gram-negative bacteremia, *Dis. Mon.,* December 1973, 1.
147. **Sanford, J. P. and Pierce, A. K.,** Current infection problems – respiratory, in *Proc. of the Int. Conf. Nosocomial Infections,* American Hospital Association, Chicago, 1971, 77.
148. **Tillotson, J. R. and Lerner, A. M.,** Pneumonias caused by gram negative bacilli, *Medicine,* 45, 65, 1966.
149. **Tillotson, J. R. and Lerner, A. M.,** Characteristics of pneumonias caused by *Escherichia coli, N. Engl. J. Med.,* 277, 115, 1967.
150. **Tillotson, J. R. and Lerner, A. M.,** Characteristics of pneumonias caused by *Bacillus proteus, Ann. Intern. Med.,* 68, 287, 1968.
151. **Tillotson, J. R. and Lerner, A. M.,** Characteristics of nonbacteremic Pseudomonas pneumonia, *Ann. Intern. Med.,* 68, 295, 1968.
152. **Tillotson, J. R. and Lerner, A. M.,** Bacteroides pneumonias: characteristics of cases with empyema, *Ann. Intern. Med.,* 68, 308, 1968.
153. **Edmondson, E. B., Reinarz, J. A., Pierce, A. K., and Sanford, J. P.,** Nebulization equipment: a potential source of infection in gram-negative pneumonias, *Am. J. Dis. Child.,* 111, 357, 1966.
154. **Reinarz, J. A., Pierce, A. K., Mays, B. B., and Sanford, J. P.,** The potential role of inhalation therapy equipment in nosocomial pulmonary infection, *J. Clin. Invest.,* 44, 831, 1965.
155. **Schulze, T., Edmondson, E. B., Pierce, A. K., and Sanford, J. P.,** Studies of a new humidifying device as a potential source of bacterial aerosols, *Am. Rev. Respir. Dis.,* 96, 517, 1967.
156. **Atkinson, W. J.,** Posture of the unconscious patient, *Lancet,* 1, 404, 1970.
157. **Cameron, J. L. and Zuidema, G. D.,** Aspiration pneumonia: magnitude and frequency of the problem, *JAMA,* 219, 1194, 1972.
158. **Heroy, W. W.,** Unrecognized aspiration (editorial), *Ann. Thorac. Surg.,* 8, 580, 1969.
159. **Mrazek, S. A.,** Bronchopneumonia in terminally ill patients, *J. Am. Geriatr. Soc.,* 17, 969, 1969.
160. **DuPont, H. L. and Spink, W. W.,** Infections due to gram-negative organisms: an analysis of 860 patients with bacteremia at the University of Minnesota Medical Center, 1958–1966, *Medicine,* 48, 307, 1969.
161. **Rogers, D. E.,** The changing pattern of life-threatening microbial disease, *N. Engl. J. Med.,* 261, 677, 1959.
162. **McCabe, W. R. and Jackson, G. G.,** Gram-negative bacteremia. I. Etiology and ecology. II. Clinical, laboratory, and therapeutic observations, *Arch. Intern. Med.,* 110, 847, 1962.
163. **Lewis, J. and Fekety, F. R., Jr.,** Gram-negative bacteremia, *Johns Hopkins Med. J.,* 124, 106, 1969.
164. **Waisbren, B. A.,** Bacteremia due to gram-negative bacilli other than the *Salmonella:* a clinical and therapeutic study, *Arch. Intern. Med.,* 88, 467, 1951.
165. **McHenry, M. C., Martin, W. J., and Wellman, W. E.,** Bacteremia due to Gram-negative bacilli: review of 113 cases encountered in the five-year period 1955 through 1959, *Ann. Intern. Med.,* 56, 207, 1962.
166. **Maiztegui, J. I., Biegeleisen, J. Z., Jr., Cherry, W. B., and Kass, E. H.,** Bacteremia due to gram-negative rods: a clinical, bacteriologic, serologic and immunofluorescent study, *N. Engl. J. Med.,* 272, 222, 1965.
167. **Hodgin, U. G. and Sanford, J. P.,** Gram-negative rod bacteremia: an analysis of 100 patients, *Am. J. Med.,* 39, 952, 1965.
168. **Freid, M. A. and Vosti, K. L.,** The importance of underlying disease in patients with gram-negative bacteremia, *Arch. Intern. Med.,* 121, 418, 1968.
169. **Altemeier, W. A., Todd, J. C., and Inge, W. W.,** Gram-negative septicemia: a growing threat, *Ann. Surg.,* 166, 530, 1967.
170. **Burton, R. C.,** Gram-negative systemic bacteraemia in alimentary tract surgery, *Med. J. Aust.,* 1, 1114, 1971.
171. **McHenry, M. C., Turnbull, R. B., Jr., Weakley, F. L., and Hawk, W. A.,** Septicemia in surgical patients with intestinal diseases, *Dis. Colon Rectum,* 14, 195, 1971.
172. **Hermans, P. E. and Washington, J. A., II,** Polymicrobial bacteremia, *Ann. Intern. Med.,* 73, 387, 1970.
173. **Wilson, W. R., Martin, W. J., Wilkowske, C. J., and Washington, J. A., II,** Anaerobic bacteremia, *Mayo Clin. Proc.,* 47, 639, 1972.
174. **Gorbach, S. L. and Bartlett, J. G.,** Anaerobic infections, *N. Engl. J. Med.,* 290, 1177, 1974.
175. **Sonnenwirth, A. C.,** Incidence of intestinal anaerobes in blood cultures, in *Anaerobic Bacteria: Role in Disease,* Balows, A., DeHaan, R. M., Dowell, V. R., Jr., and Guze, L. B., Eds., Charles C Thomas, Springfield, Ill., 1974, 157.
176. **Ellner, P. D., Granato, P. A., and May, C. B.,** Recovery and identification of anaerobes: a system suitable for the routine clinical laboratory, *Appl. Microbiol.,* 26, 904, 1973.
177. **Felner, J. M. and Dowell, V. R., Jr.,** "Bacteroides" bacteremia, *Am. J. Med.,* 50, 787, 1971.

178. Marcoux, J. A., Zabransky, R. J., Washington, J. A., II, Wellman, W. E., and Martin, W. J., Bacteroides bacteremia, *Minn. Med.*, 53, 1169, 1970.
179. Alpern, R. J. and Dowell, V. R., Jr., *Clostridium septicum* infections and malignancy, *JAMA*, 209, 385, 1969.
180. Myerowitz, R. L., Medeiros, A. A., and O'Brien, T. F., Recent experience with bacillemia due to gram-negative organisms, *J. Infect. Dis.*, 124, 239, 1971.
181. Conn, H. O. and Fessel, J. M., Spontaneous bacterial peritonitis in cirrhosis: variations on a theme, *Medicine*, 50, 161, 1971.
182. Correia, J. P. and Conn, H. O., Spontaneous bacterial peritonitis in cirrhosis: endemic or epidemic?, *Med. Clin. North Am.*, 59, 963, 1975.
183. Conn, H. O., Spontaneous bacterial peritonitis: multiple revisitations (editorial), *Gastroenterology*, 70, 455, 1976.
184. DeMeo, A. N. and Andersen, B. R., Defective chemotaxis associated with a serum inhibitor in cirrhotic patients, *N. Engl. J. Med.*, 286, 735, 1972.
185. Targan, S. R., Chow, A. W., and Guze, L. B., Role of anaerobic bacteria in spontaneous peritonitis of cirrhosis: report of two cases and review of the literature, *Am. J. Med.*, 62, 397, 1977.
186. Slaughter, L., Morris, J. E., and Starr, A., Prosthetic valvular endocarditis: a 12-year review, *Circulation*, 47, 1319, 1973.
187. Dismukes, W. E., Karchmer, A. W., Buckley, M. J., Austen, W. G., and Swartz, M. N., Prosthetic valve endocarditis: analysis of 38 cases, *Circulation*, 48, 365, 1973.
188. Wilson, W. R., Jaumin, P. M., Danielson, G. K., Giuliani, E. R., Washington, J. A., II, and Geraci, J. E., Prosthetic valve endocarditis, *Ann. Intern. Med.*, 82, 751, 1975.
189. Blakemore, W. S., McGarrity, G. J., Thurer, R. J., MacVaugh, H., III, and Coriell, L. L., Infection by air-borne bacteria with cardiopulmonary bypass, *Surgery*, 70, 830, 1971.
190. Ankeney, J. L. and Parker, R. F., Staphylococcal endocarditis following open heart surgery related to positive intraoperative blood cultures, in *Prosthetic Heart Valves*, Brewer, L. A., III, Cooley, D. A., Davila, J. C., Merendino, K. A., and Sirak, H. D., Eds., Charles C Thomas, Springfield, Ill., 1969, 719.
191. Block, P. C., DeSanctis, R. W., Weinberg, A. N., and Austen, W. G., Prosthetic valve endocarditis, *J. Thorac. Cardiovasc. Surg.*, 60, 540, 1970.
192. Okies, J. E., Viroslav, J., and Williams, T. W., Jr., Endocarditis after cardiac valvular replacement, *Chest*, 59, 198, 1971.
193. Fraser, R. S., Rossall, R. E., and Dvorkin, J., Bacterial endocarditis occurring after open-heart surgery, *Can. Med. Assoc. J.*, 96, 1551, 1967.
194. Shafer, R. B. and Hall, W. H., Bacterial endocarditis following open heart surgery, *Am. J. Cardiol.*, 25, 602, 1970.
195. Davis, A., Binder, M. J., and Finegold, S. M., Late infection in patients with Starr-Edwards prosthetic cardiac valves, *Antimicrob. Agents Chemother.*, 1965, 97, 1966.
196. Cohn, L. H., Roberts, W. C., Rockoff, S. D., and Morrow, A. G., Bacterial endocarditis following aortic valve replacement: clinical and pathologic correlations, *Circulation*, 33, 209, 1966.
197. Amoury, R. A., Bowman, F. O., Jr., and Malm, J. R., Endocarditis associated with intracardiac prostheses: diagnosis, management, and prophylaxis, *J. Thorac. Cardiovasc. Surg.*, 51, 36, 1966.
198. Weinstein, L., Infected prosthetic valves: a diagnostic and therapeutic dilemma (editorial), *N. Engl. J. Med.*, 286, 1108, 1972.
199. Wilson, W. R., Prosthetic valve endocarditis: incidence, antamoic location, cause, morbidity, mortality, in *Infections of Prosthetic Heart Valves and Vascular Grafts*, Duma, R. J., Ed., University Park Press, Baltimore, 1977, 3.
200. Liekweg, W. G., Levinson, S. A., and Greenfield, L. S., Infections of vascular grafts: incidence, anatomic location, etiologic agents, morbidity, and mortality, in *Infections of Prosthetic Heart Valves and Vascular Grafts*, Duma, R. J., Ed., University Park Press, Baltimore, 1977, 239.
201. Bouhoutsos, J., Chavatzas, D., Martin, P., and Morris, T., Infected synthetic arterial grafts, *Br. J. Surg.*, 61, 108, 1974.
202. Carter, S. C., Cohen, A., and Whelan, T. J., Clinical experience with management of the infected dacron graft, *Ann. Surg.*, 158, 249, 1963.
203. Conn, J. H., Hardy, J. D., Chavez, C. M., and Fain, W. R., Infected arterial grafts: experience in 22 cases with emphasis on unusual bacteria and technics, *Ann. Surg.*, 171, 704, 1970.
204. Diethrich, E. B., Noon, G. P., Liddicoat, J. E., and De Bakey, M. E., Treatment of infected aortofemoral arterial prosthesis, *Surgery*, 68, 1044, 1970.
205. Fry, W. J. and Lindenauer, S. M., Infection complicating the use of plastic arterial implants, *Arch. Surg.*, 94, 600, 1967.
206. Goldstone, J. and Moore, W. S., Infection in vascular prostheses: clinical manifestations and surgical management, *Am. J. Surg.*, 128, 225, 1974.
207. Hoffert, P. W., Gensler, S., and Haimovici, H., Infection complicating arterial grafts: personal experience with 12 cases and review of the literature, *Arch. Surg.*, 90, 427, 1965.
208. Najafi, H., Javid, H., Dye, W. S., Hunter, J. A., and Julian, O. C., Management of infected arterial implants, *Surgery*, 65, 539, 1969.
209. Schramel, R. J. and Creech, O., Jr., Effects of infection and exposure on synthetic arterial prostheses, *Arch. Surg.*, 78, 271, 1959.

210. **Shaw, R. S. and Baue, A. E.,** Management of sepsis complicating arterial reconstructive surgery, *Surgery,* 53, 75, 1963.
211. **Smith, R. B., III, Lowry, K., and Perdue, G. D.,** Management of the infected arterial prosthesis in the lower extremity, *Am. Surg.,* 33, 711, 1967.
212. **Smith, R. F. and Szilagyi, D. E.,** Healing complications with plastic arterial implants, *Arch. Surg.,* 82, 14, 1961.
213. **Szilagyi, D. E., Smith, R. F., Elliott, J. P., and Vrandecic, M. P.,** Infection in arterial reconstruction with synthetic grafts, *Ann. Surg.,* 176, 321, 1972.
214. **Willwerth, B. M. and Waldhausen, J. A.,** Infection of arterial prostheses, *Surg. Gynecol. Obstet.,* 139, 446, 1974.
215. **Van De Water, J. M. and Gaal, P. G.,** Management of patients with infected vascular prostheses, *Am. Surg.,* 31, 651, 1965.
216. **Corman, L. C. and Levison, M. E.,** Sustained bacteremia and transvenous cardiac pacemakers, *JAMA,* 233, 264, 1975.
217. **Lemire, G. G., Morin, J. E., and Dobell, A. R. C.,** Pacemaker infections: a 12-year review, *Can. J. Surg.,* 18, 181, 1975.
218. **Svanbom, M., Gästrin, B., and Rodriguez, L.,** Transvenous cardiac pacemaker as a focus of salmonella infection in a patient with heart block, *Acta Med. Scand.,* 196, 281, 1974.
219. **Ma, P., Delaney, W. E., and Grace, W. J.,** Incidence of septicemia in patients with cardiac pacemakers, *Crit. Care Med.,* 2, 135, 1974.
220. **Sedaghat, A.,** Permanent transvenous pacemaker infection with septicemia, *N.Y. State J. Med.,* 74, 868, 1974.
221. **Klevan, D., Zinneman, H. H., Hall, W. H., and Sako, Y.,** Contaminated pacemaker lead wire: causing chronic pseudomonas septicemia, *Minn. Med.,* 56, 750, 1973.
222. **Schoenbaum, S. C., Gardner, P., and Shillito, J.,** Infections of cerebrospinal fluid shunts: epidemiology clinical manifestations, and therapy, *J. Infect. Dis.,* 131, 543, 1975.
223. **Yashon, D. and Sugar, O.,** Today's problems in hydrocephalus, *Arch. Dis. Child.,* 39, 58, 1964.
224. **Anderson, F. M.,** Ventriculo-auriculostomy in treatment of hydrocephalus, *J. Neurosurg.,* 16, 551, 1959.
225. **Schimke, R. T., Black, P. H., Mark, V. H., and Swartz, M. N.,** Indolent *Staphylococcus albus* or *aureus* bacteremia after ventriculoatriostomy: role of foreign body in its initiation and perpetuation, *N. Engl. J. Med.,* 264, 264, 1961.
226. **Bruce, A. M., Lorber, J., Shedden, W. I. H., and Zachary, R. B.,** Persistent bacteraemia following ventriculo-caval shunt operations for hydrocephalus in infants, *Dev. Med. Child Neurol.,* 5, 461, 1963.
227. **Scarff, J. E.,** Treatment of hydrocephalus: an historical and critical review of methods and results, *J. Neurol. Neurosurg. Psychiatry,* 26, 1, 1963.
228. **Overton, M. C., III and Snodgrass, S. R.,** Ventriculo-venous shunts for infantile hydrocephalus: a review of five years' experience with this method, *J. Neurosurg.,* 23, 517, 1965.
229. **Eckstein, H. B. and Cooper, D. G. W.,** The complications of ventriculo-atrial shunts with the Holter valve for hydrocephalus, *Z. Kinderchir.,* 5, 309, 1968.
230. **Forrest, D. M. and Cooper, D. G. W.,** Complications of ventriculo-atrial shunts: a review of 455 cases, *J. Neurosurg.,* 29, 506, 1968.
231. **Luthardt, T.,** Bacterial infections in ventriculo-auricular shunt systems, *Dev. Med. Child Neurol.,* 12 (Suppl. 22), 105, 1970.
232. **Illingworth, R. D., Logue, V., Symon, L., and Uemura, K.,** The ventriculocaval shunt in the treatment of adult hydrocephalus: results and complications in 101 patients, *J. Neurosurg.,* 35, 681, 1971.
233. **McLaurin, R. L. and Dodson, D.,** Infected ventriculoatrial shunts: some principles of treatment, *Dev. Med. Child Neurol.,* 13 (Suppl. 25), 71, 1971.
234. **Shurtleff, D. B., Christie, D., and Foltz, E. L.,** Ventriculoauriculostomy-associated infection: a 12-year study, *J. Neurosurg.,* 35, 686, 1971.
235. **Villani, R., Paoletti, P., and Gaini, S. M.,** Experience with ventriculo-peritoneal shunts, *Dev. Med. Child Neurol.,* 13 (Suppl. 25), 101, 1971.
236. **Little, J. R., Rhoton, A. L., Jr., and Mellinger, J. F.,** Comparison of ventriculoperitoneal and ventriculoatrial shunts for hydrocephalus in children, *Mayo Clin. Proc.,* 47, 396, 1972.
237. **Anderson, F. M.,** Ventriculocardiac shunts: identification and control of practical problems in 143 cases, *J. Pediatr.,* 82, 222, 1973.
238. **Robertson, J. S., Maraqa, M. I., and Jennett, B.,** Ventriculoperitoneal shunting for hydrocephalus, *Br. Med. J.,* 2, 289, 1973.
239. **Hassen, A.,** Gram-negative bacteremic shock, *Med. Clin. North Am.,* 57, 1403, 1973.
240. **Mencher, W. H. and Leiter, H. E.,** Anaerobic infections following operations on the urinary tract, *Surg. Gynecol. Obstet.,* 66, 677, 1938.
241. **Munroe, D. S. and Cockcroft, W. H.,** Septicaemia due to gram-negative bacilli, *Can. Med. Assoc. J.,* 72, 586, 1955.
242. **Martin, C. M. and Bookrajian, E. N.,** Bacteriuria prevention after indwelling urinary catheterization: a controlled study, *Arch. Intern. Med.,* 110, 703, 1962.
243. **Swenson, R. M., Michaelson, T. C., Daly, M. J., and Spaulding, E. H.,** Anaerobic bacterial infections of the female genital tract, *Obstet. Gynecol.,* 42, 538, 1973.
244. **Thadepalli, H., Gorbach, S. L., and Keith, L.,** Anaerobic infections of the female genital tract: bacteriologic and therapeutic aspects, *Am. J. Obstet. Gynecol.,* 117, 1034, 1973.

245. **Pearson, H. E. and Anderson, G. V.,** Genital bacteroidal abscesses in women, *Am. J. Obstet. Gynecol.*, 107, 1264, 1970.
246. **Ramsay, A. M.,** The significance of *Clostridium welchii* in the cervical swab and blood-stream in postpartum and postabortum sepsis, *J. Obstet. Gynaecol. Br. Emp.*, 56, 247, 1949.
247. **Lemierre, A. and Reilly, J.,** Les bactériémies a "Bacillus ramosus," *Presse Med.*, 1, 385, 1938.
248. **Hill, R. B., Jr., Rowlands, D. T., Jr., and Rifkind, D.,** Infectious pulmonary disease in patients receiving immunosuppressive therapy for organ transplantation, *N. Engl. J. Med.*, 271, 1021, 1964.
249. **McHenry, M. C., Martin, W. J., Hargraves, M. M., and Baggenstoss, A. H.,** Bacteremia in patients with neoplastic or hematologic disease, *Mayo Clin. Proc.*, 37, 43, 1962.
250. **Hill, R. B., Jr., Dahrling, B. E., II, Starzl, T. E., and Rifkind, D.,** Death after transplantation: an analysis of sixty cases (editorial), *Am. J. Med.*, 42, 327, 1967.
251. **Bodey, G. P., Buckley, M., Sathe, Y. S., and Freireich, E. J.,** Quantitative relationships betweem circulating leukocytes and infection in patients with acute leukemia, *Ann. Intern. Med.*, 64, 328, 1966.
252. **Krick, J. A. and Remington, J. S.,** Opportunistic invasive fungal infections in patients with leukaemia and lymphoma, *Clin. Haematol.*, 5, 249, 1976.
253. **Isiadinso, O. A.,** *Listeria* sepsis and meningitis: a complication of renal transplantation, *JAMA*, 234, 842, 1975.
254. **Louria, D. B., Hensle, T., Armstrong, D., Collins, H. S., Blevins, A., Krugman, D., and Buse, M.,** Listeriosis complicating malignant disease: a new association, *Ann. Intern. Med.*, 67, 261, 1967.
255. **Donaldson, S. S., Moore, M. R., Rosenberg, S. A., and Vosti, K. L.,** Characterization of postsplenectomy bacteremia among patients with and without lymphoma, *N. Engl. J. Med.*, 287, 69, 1972.
256. **Chilcote, R. R., Baehner, R. L., Hammond, D., and the Investigators and Special Studies Committee of the Children's Cancer Study Group,** Septicemia and meningitis in children splenectomized for Hodgkin's disease, *N. Engl. J. Med.*, 295, 798, 1976.
257. **Winslow, E. J., Loeb, H. S., Rahimtoola, S. H., Kamath, S., and Gunnar, R. M.,** Hemodynamic studies and results of therapy in 50 patients with bacteremic shock, *Am. J. Med.*, 54, 421, 1973.
258. **Nishijima, H., Weil, M. H., Shubin, H., and Cavanilles, J.,** Hemodynamic and metabolic studies on shock associated with gram negative bacteremia, *Medicine*, 52, 287, 1973.
259. **MacLean, L. D., Mulligan, W. G., McLean, A. P. H., and Duff, J. H.,** Patterns of septic shock in man: a detailed study of 56 patients, *Ann. Surg.*, 166, 543, 1967.
260. **Christy, J. H.,** Pathophysiology of gram-negative shock, *Am. Heart J.*, 81, 694, 1971.
261. **Lightfoot, R. W., Jr. and Gotschlich, E. C.,** Gonococcal disease, *Am. J. Med.*, 56, 347, 1974.
262. **Holmes, K. K., Counts, G. W., and Beaty, H. N.,** Disseminated gonococcal infection, *Ann. Intern. Med.*, 74, 979, 1971.
263. **Ketover, B. P., Young, L. S., and Armstrong, D.,** Septicemia due to *Aeromonas hydrophila:* clinical and immunologic aspects, *J. Infect. Dis.*, 127, 284, 1973.
264. **Reynolds, H. Y., Levine, A. S., Wood, R. E., Zierdt, C. H., Dale, D. C., and Pennington, J. E.,** *Pseudomonas aeruginosa* infections: persisting problems and current research to find new therapies, *Ann. Intern. Med.*, 82, 819, 1975.
265. **Roberts, G. D., Brewer, N. S., and Hermans, P. E.,** Diagnosis of nocardiosis by blood culture, *Mayo Clin. Proc.*, 49, 293, 1974.
266. **Balandran, L., Rothschild, H., Pugh, N., and Seabury, J.,** A cutaneous manifestation of systemic candidiasis, *Ann. Intern. Med.*, 78, 400, 1973.
267. **Ratnoff, O. D. and Nebehay, W. G.,** Multiple coagulative defects in a patient with the Waterhouse-Friderichsen syndrome, *Ann. Intern. Med.*, 56, 627, 1962.
268. **McGehee, W. G., Rapaport, S. I., and Hjort, P. F.,** Intravascular coagulation in fulminant meningococcemia, *Ann. Intern. Med.*, 67, 250, 1967.
269. **Corrigan, J. J., Jr., Ray, W. L., and May, N.,** Changes in the blood coagulation system associated with septicemia, *N. Engl. J. Med.*, 279, 851, 1968.
270. **Argen, R. J., Wilson, C. H., Jr., and Wood, P.,** Suppurative arthritis: clinical features of 42 cases, *Arch. Intern. Med.*, 117, 661, 1966.
271. **Waldvogel, F. A., Medoff, G., and Swartz, M. N.,** Osteomyelitis: a review of clinical features, therapeutic considerations and unusual aspects, *N. Engl. J. Med.*, 282, 198; 260; 316, 1970.
272. **Robin, E. D., Carey, L. C., Grenvik, A., Glauser, F., and Gaudio, R.,** Capillary leak syndrome with pulmonary edema, *Arch. Intern. Med.*, 130, 66, 1972.
273. **Riordan, J. F. and Walters, G.,** Pulmonary oedema in bacterial shock, *Lancet*, 1, 719, 1968.
274. **Snell, J. D., Jr. and Ramsey, L. H.,** Pulmonary edema as a result of endotoxemia, *Am. J. Physiol.*, 217, 170, 1969.
275. **Wilson, W. R. and Washington, J. A., II,** Infective endocarditis: a changing spectrum (editorial), *Mayo Clin. Proc.*, 54, 254, 1977.
276. **Lerner, P. I. and Weinstein, L.,** Infective endocarditis in the antibiotic era, *N. Engl. J. Med.*, 274, 199; 259; 323; 388, 1966.
277. **Kaye, D.,** Infecting microorganisms, in *Infective Endocarditis*, Kaye, D., Ed., University Park Press, Baltimore, 1976, 43.
278. **Tompsett, R.,** Bacterial endocarditis: changes in the clinical spectrum, *Arch. Intern. Med.*, 119, 329, 1967.

279. **Shinebourne, E. A., Cripps, C. M., Hayward, G. W., and Shooter, R. A.,** Bacterial endocarditis 1956–1965: analysis of clinical features and treatment in relation to prognosis and mortality, *Br. Heart J.,* 31, 536, 1969.
280. **Wilson, L. M.,** Etiology of bacterial endocarditis: before and since the introduction of antibiotics, *Ann. Intern. Med.,* 58, 946, 1963.
281. **Geraci, J. E., Wilkowske, C. J., Wilson, W. R., and Washington, J. A., II,** *Haemophilus* endocarditis: report of 14 patients, *Mayo Clin. Proc.,* 52, 209, 1977.
282. **Mills, J. and Drew, D.,** *Serratia marcescens* endocarditis: a regional illness associated with intravenous drug abuse, *Ann. Intern. Med.,* 84, 29, 1976.
283. **Reyes, M. P., Palutke, W. A., Wylin, R. F., and Lerner, A. M.,** Pseudomonas endocarditis in the Detroit Medical Center, 1969–1972, *Medicine,* 52, 173, 1973.
284. **Roberts, W. C. and Buchbinder, N. A.,** Right-sided valvular infective endocarditis: a clinicopathologic study of twelve necropsy patients, *Am. J. Med.,* 53, 7, 1972.
285. **Cherubin, C. E. and Neu, H. C.,** Infective endocarditis at the Presbyterian Hospital in New York City from 1938–1967, *Am. J. Med.,* 51, 83, 1971.
286. **Weinstein, L. and Rubin, R. H.,** Infective endocarditis – 1973, *Prog. Cardiovasc. Dis.,* 16, 239, 1973.
287. **Banks, T., Fletcher, R., and Ali, N.,** Infective endocarditis in heroin addicts, *Am. J. Med.,* 55, 444, 1973.
288. **Nolan, C. M. and Beaty, H. N.,** *Staphylococcus aureus* bacteremia: current clinical patterns, *Am. J. Med.,* 60, 495, 1976.
289. **Cluff, L. E., Reynolds, R. C., Page, D. L., and Breckenridge, J. L.,** Staphylococcal bacteremia and altered host resistance, *Ann. Intern. Med.,* 69, 859, 1968.
290. **Young, R. C., Bennett, J. E., Geelhoed, G. W., and Levine, A. S.,** Fungemia with compromised host resistance: a study of 70 cases, *Ann. Intern. Med.,* 80, 605, 1974.
291. **Weinstein, A. J., Johnson, E. H., and Moellering, R. C., Jr.,** *Candida* endophthalmitis: a complication of candidemia, *Arch. Intern. Med.,* 132, 749, 1973.

Chapter 1
SEPSIS

ORGANISMS FROM BLOOD CULTURES AT THE MAYO CLINIC: 1968 TO 1975

Duane M. Ilstrup

INTRODUCTION

Since 1968, blood culture organisms have routinely been data banked at the Mayo Clinic using the data collection sheet presented in another section in this chapter. With these data, we can look at the broad scope of sepsis over time in a large medical institution. We will first examine a series of 15 organism groups to determine if the incidence of these groups has changed over the years. We will also look, in detail, at the year 1975 and enumerate all of the organisms that were isolated in that year. Finally, for the year 1975, we will examine the distributions of the time in days from the date the specimen was taken to the date the organism was detected for each of the 15 organism groups mentioned above.

ORGANISMS SEEN OVER TIME

When trying to describe the distribution of organisms seen in an institution, it is good practice to differentiate between specimens and patients. Many cultures are sometimes taken from the same patient and if this fact is not taken into account, the percentage of positive cultures with a given organism might be extremely influenced by one or more patients from whom a large number of cultures were taken. In the discussion that follows, the denominator of any percentage given will be the total number of unique patients seen with positive cultures in that year. The data for the 15 organism groups chosen to be studied appear in Table 1. It should be noted that the totals in any given year may sum to more than 100% because some patient's cultures yielded more than one organism.

This table also shows that the proportion of all positive patients has increased over the years for *Corynebacterium* (including *Propionibacterium*), *Serratia,* and *Staphylococcus epidermidis,* and the proportion of all positive patients has decreased for *Klebsiella, Streptococcus* group D, viridans *Streptococcus, Bacteroides,* and *Pseudomonas aeruginosa.* The remaining seven organism groups exhibit no meaningful trends.

1975 ORGANISMS

In 1975 about 25,000 blood culture specimens from approximately 7000 patients were received by the Section of Clinical Microbiology. Of these, there were 2409 positive specimens from 1211 patients. The numbers and percentages of the positive patients for organisms seen in 1975 are presented in Table 2. The most frequently seen organism at the Mayo Clinic is *Corynebacterium* (including *Propionibacterium*). Thirty percent of the positive patients had *Corynebacterium* isolated from their cultures. The next most frequent organism was *Staphylococcus epidermidis,* 19.5%, followed closely by *Escherichia coli* with 16.7%. The only other organism with a major share of the positive patients was *Staphylococcus aureus* with 9.8%.

TIME TO POSITIVITY

The time in days to initial detection of positivity for cultures in 1975 is shown in Table 3. For example, the minimum number of days required for *Corynebacterium* to appear in a culture was 1 day, the maximum time was 17 days; the median time (time such that 50% of the cultures appeared before and 50% after that time) was 8 days; the average time was 9.5 days; and the 90th percentile (time such that 90% of the cultures appeared before that time) was 14 days.

If one is interested in determining a "typical" time to positivity, the median is a much better estimate than the average, which is usually skewed to high values of days to positivity and is, therefore, not a good estimate of a typical time. The median, however, is a much better-behaved statistic and will be a good estimate of what would be expected of a typical culture.

Many times a clinician may want to know how long he should reasonably wait to determine if a

TABLE 1

Percent of Positive Patients by Organism Group

Organism	Year 1968	1969	1970	1971	1972	1973	1974	1975
Corynebacterium	22.3	17.4	23.1	30.6	33.5	32.7	25.8	30.2
Escherichia coli	13.7	13.3	14.9	18.3	16.3	14.7	13.1	16.7
Klebsiella	7.4	8.0	6.6	4.8	5.7	5.8	5.6	4.5
Enterobacter	2.0	3.3	1.9	1.6	1.6	2.1	1.6	2.1
Serratia	1.4	0.6	2.3	1.5	0.6	1.3	2.4	2.9
Proteus	1.1	3.3	3.8	3.1	2.9	2.9	3.0	2.4
Haemophilus	0.3	1.2	1.5	1.7	1.5	1.8	1.2	1.5
Streptococcus pneumoniae	2.0	2.0	2.1	1.9	2.2	2.4	2.8	2.7
Streptococcus group D	6.9	5.2	4.8	2.2	1.6	2.3	2.1	3.4
Streptococcus viridans group	7.1	8.5	7.8	8.9	6.7	3.9	6.3	4.7
Bacteroides	7.1	8.4	8.9	6.8	8.4	6.4	5.7	5.2
Staphylococcus aureus	10.9	7.8	9.5	7.6	8.3	9.5	9.3	9.8
Staphylococcus epidermidis	12.6	12.9	14.4	11.9	11.3	15.7	19.1	19.5
Pseudomonas aeruginosa	6.3	6.4	7.1	4.3	3.4	4.8	4.1	4.5
Candida	0.6	0.4	0.8	1.6	0.8	0.9	0.6	1.5

TABLE 2

Number and Percentage of Positive Patients by Organism Group for 1975

Organism	Number of patients	Positive patients (%)[a]	Organism	Number of patients	Positive patients (%)
Unspecified	6	0.5	*Streptococcus,* beta hemolytic group F	1	0.1
Bacillus	19	1.6	*Streptococcus,* beta hemolytic, other specified groups	1	0.1
Clostridium	17	1.4	*Eubacterium*	2	0.2
Corynebacterium	366	30.2	Non-fermenting Gram-negative bacilli	38	3.1
Escherichia coli	202	16.7	*Neisseria*	3	0.2
Salmonella	1	0.1	*Bacteroides*	63	5.2
Citrobacter	9	0.7	*Micrococcus*	2	0.2
Klebsiella	54	4.5	*Staphylococcus aureus*	119	9.8
Enterobacter	25	2.1	*Staphylococcus epidermidis*	236	19.5
Serratia	35	2.9	*Peptostreptococcus*	9	0.7
Proteus	29	2.4	*Peptococcus*	1	0.1
Providencia	1	0.1	*Veillonella*	3	0.2
Haemophilus	18	1.5	*Pseudomonas aeruginosa*	55	4.5
Streptococcus pneumoniae	33	2.7	*Campylobacter*	1	0.1
Listeria monocytogenes	1	0.1	*Aeromonas*	3	0.2
Streptococcus – unspecified	14	1.2	*Candida*	18	1.5
Streptococcus, group D (enterococcus)	41	3.4	*Cryptococcus*	1	0.1
Streptococcus, viridans	57	4.7	*Lactobacillus*	2	0.2
Streptococcus, beta hemolytic group A	8	0.7	Other	40	3.3
Streptococcus, beta hemolytic group B	3	0.2			

[a]Total number of positive patients = 1211.

TABLE 3

Time in Days to Positivity by Organism Group for 1975

Organism	Minimum	Maximum	Median	Average	90th Percentile
Corynebacterium	1	17	8.0	9.5	14.0
Escherichia coli	1	7	1.0	1.7	3.0
Klebsiella	1	15	1.0	2.3	4.0
Enterobacter	1	5	1.0	1.7	3.0
Serratia	1	5	1.0	1.5	2.0
Proteus	1	9	2.0	2.3	3.0
Haemophilus	1	7	2.0	2.9	7.0
Streptococcus pneumoniae	1	4	1.0	1.5	2.0
Streptococcus, group D	1	9	2.0	2.1	3.0
Streptococcus, viridans group	1	16	2.0	2.6	4.0
Bacteroides	1	13	3.0	4.0	7.0
Staphylococcus aureus	1	13	2.0	3.3	7.0
Staphylococcus epidermidis	1	16	3.0	4.2	6.0
Pseudomonas aeruginosa	1	14	2.0	3.0	6.0
Candida	1	14	5.0	6.3	13.0

patient has a positive culture. For this reason, the maximum observed time to positivity and the 90th percentile were included in Table 3.

The maximum is the longest time that we observed for a culture to become positive with a given organism. The maximum is a function of sample size, i.e. the larger the sample size, the larger the maximum is; therefore, the maximums presented in Table 3 are not the largest values possible; they are only the largest observed values.

The 90th percentiles have been included to give an estimate of the time required for most of the cultures to become positive. In other words, 9 out of 10 positive cultures would have become positive by the quoted times. This statistic is, again, dependent on sample size but should, regardless of that fact, be a useful estimate.

Chapter 1

SEPSIS

CHARACTERISTICS OF BACTEREMIA RELEVANT TO ITS LABORATORY DIAGNOSIS

John A. Washington II

Bacteremias of clinical significance are generally continuous or intermittent in type.[1] Those associated with the first 2 or 3 weeks of typhoid fever, brucellosis, and endocarditis or other intravascular infection are usually continuous.

In studies of six patients with bacterial endocarditis conducted by Beeson et al.,[2] it was demonstrated that the rate of discharge of bacteria was rather constant and quite uniform in its order of magnitude. Similar findings had previously been reported by Weiss and Ottenberg,[3] who studied four cases with subacute bacterial endocarditis and found the order of magnitude of bacteremia to have been relatively constant in each case; however, they also found that in three of the cases there were sudden showers of larger numbers of bacteria, unrelated to the severity of the disease but temporally related to the patients' temperatures. Such intermittent showers of bacteria were not observed in any of the patients studied by Beeson et al.,[2] but were noted by Wright,[4] who was unable to document their temporal association with fever. Based on their findings, however, Weiss and Ottenberg[3] suggested that more positive results and higher bacterial counts would be obtained if blood cultures were collected just prior to an anticipated rise in temperature.

Studies were made by Beeson et al.[2] and by Mallén et al.[5] to determine whether cultures of arterial blood would be more likely to be positive than those of venous blood in patients with bacterial endocarditis. In both instances, it was found that cultures of venous blood were just as likely to be positive as those of arterial blood. It seems unlikely, however, that these findings are applicable to fungal endocarditis in which cultures of venous blood are frequently negative, especially in those cases caused by *Aspergillus.*[6-9] Whether arterial blood cultures, which have been shown to have greater diagnostic value than venous blood cultures in *Candida* sepsis,[10] should be conducted in cases of suspected fungal endocarditis remains to be seen. Factors related to the manner in which fungal blood cultures are performed will be discussed in Chapter 3.

The order of magnitude of the bacteremia associated with endocarditis was shown by Weiss and Ottenberg[3] to range from approximately 10 colonies per milliliter (col/ml) to over 4000 col/ml. In the studies by Beeson et al.[2] Arterial counts varied considerably from patient to patient, ranging from 4 to 697 col/ml; however, as previously stated, the magnitude of the bacteremia in each patient studied remained quite uniform. In this study, blood taken from the antecubital vein gave counts which were only slightly lower than those in arterial blood. In general, the colony counts reported by Mallén et al.[5] were substantially lower than those reported in the two earlier studies. In studies reported by Werner et al.,[11] the order of magnitude of bacteremia in endocarditis was also low (Table 1). Over 50% of the cultures collected from patients with streptococcal endocarditis had only 1 to 30 col/ml, and the same low order of magnitude of bacteremia was observed in 70% of the cultures in patients with staphylococcal endocarditis. Only 17% of the quantitative blood cultures contained more than 100 col/ml, and it is likely that the actual incidence was lower, as those cases in which broth cultures only were positive could not be analyzed. The reasons for the varying colony counts in the studies by Weiss and Ottenberg,[3] Beeson et al.,[2] Mallén et al.[5] and Werner et al.,[11] are speculative but may reflect patient selection.[11]

TABLE 1

Order of Magnitude of Bacteremia in Endocarditis

	Streptococcal		Staphylococcal	
Colonies/ml	Number	%	Number	%
1–30	221	54	12	70
31–50	60	14	1	6
51–100	63	15	2	12
>100	71	17	2	12

Adapted from Werner, A. S., Cobbs, C. G., Kaye, D., and Hook, E. W., *JAMA,* 202, 199, 1967.

TABLE 2

Order of Magnitude of Bacteremia in Adults: University of Minnesota, 1958–1966

Pour plates (col/ml)	Patients positive[a]	
	Number	%
0	267	58
1–9	122	26
10–49	33	7
≥50	42	9
Total	464	100

[a]Excluding patients with polymicrobial and Bacteroides bacteremia.

Adapted from DuPont, H. L. and Spink, W. W., *Medicine,* 48, 307, 1969.

TABLE 3

Order of Magnitude of Bacteremia in Neonatal Septicemia due to *Escherichia coli*

Colonies/ml	Cultures positive	
	Number	%
1–49	19	54
50–1000	5	14
>1000	11	32
Total	35	100

Adapted from Dietzman, D. E., Fisher, G. W., and Schoenknecht, F. D., *J. Pediatr.,* 85, 128, 1974.

Although the bacteremias associated with brucellosis and typhoid fever are considered to be continuous, their duration is usually limited to the first few weeks of illness.[1] Moreover, in both of these diseases, it has been shown that cultures of the bone marrow may be more frequently positive than those of blood.[5,12–14] Conversely, Hamilton[13] demonstrated that the frequency of positivity of a single blood culture and a single bone marrow aspirate culture for brucella was identical, but was significantly higher when multiple blood cultures were performed.

In their review of Gram-negative bacteremias in 860 patients at the University of Minnesota, DuPont and Spink[15] analyzed pour plate colony counts in 464 adults (Table 2). There were 267 (58%) cases in which pour plate cultures were sterile and the broth cultures only were positive. In an additional 122 (26%) cases, there were 1 to 9 col/ml. Of the remaining cases, 33 (7%) and 42 (9%) had 10 to 49 and ≥50 col/ml, respectively. In a study of bacteremia associated with genitourinary tract manipulation, Sullivan et al.,[16] found that in nearly all instances the average number of colonies present was ≤10/ml, the exceptions being bacteremias due to *Klebsiella pneumoniae,* in which an average of 84 col/ml was present. Although the issue of the order of magnitude of bacteremia in conditions other than endocarditis has not been extensively studied, it does appear that such bacteremias in adults are of low orders of magnitude.

In studies of neonatal septicemia due to *Escherichia coli,* Dietzman et al.,[17] found that colony counts in blood varied widely from less than 5 to more than 1000 col/ml (Table 3). Slightly more than one half of the patients had colony counts of 50/ml, while a third had counts exceeding 1000/ml. However, in three of the five infants who had more than one positive culture within 2 days, the counts were quite variable and ranged in two infants from >1000 col/ml to 16 and 24 col/ml, respectively, and in one infant from 4 to 44 col/ml. Albers et al.[18] examined colony counts of bacteria in blood cultures of three infants with prolonged bacteremia and of one infant with septicemia and found counts ranging from 1 to 126 col/ml. In three of the cases, the counts progressively increased over a period of 4 to 6 days. All of the infants in this study were asymptomatic, in contrast to those investigated by Dietzman et al.;[17] this may account for the lower order of magnitude of their bacteremias.

With the possible exception of neonatal septicemia, therefore, bacteremias are generally of a low order of magnitude and are in a significant number of instances undetectable by pour plate techniques in which only 1 ml of blood is customarily cultured. Moreover, bacteremias, other than those associated with endocarditis or with the acute phases of brucellosis and typhoid fever, are ordinarily intermittent. These concepts are important to remember when attempting to establish the etiology of bacteremias by microbiological methods. It is essential, therefore, for an adequate volume of blood to be cultured and for separate cultures to be collected at intervals during a 24-hr period of time.

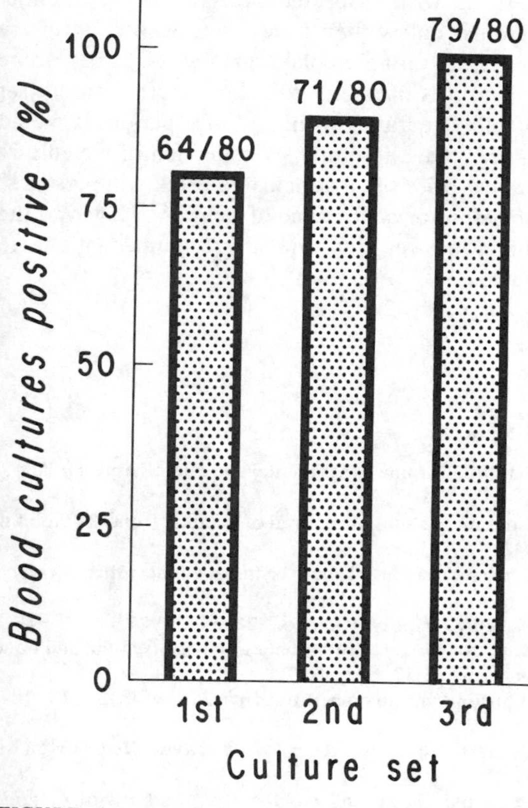

FIGURE 1. Cumulative rates of positivity in three blood culture sets from 80 bacteremic patients. (From Washington, J. A., II, *Mayo Clin. Proc.*, 50, 91, 1975. With permission.)

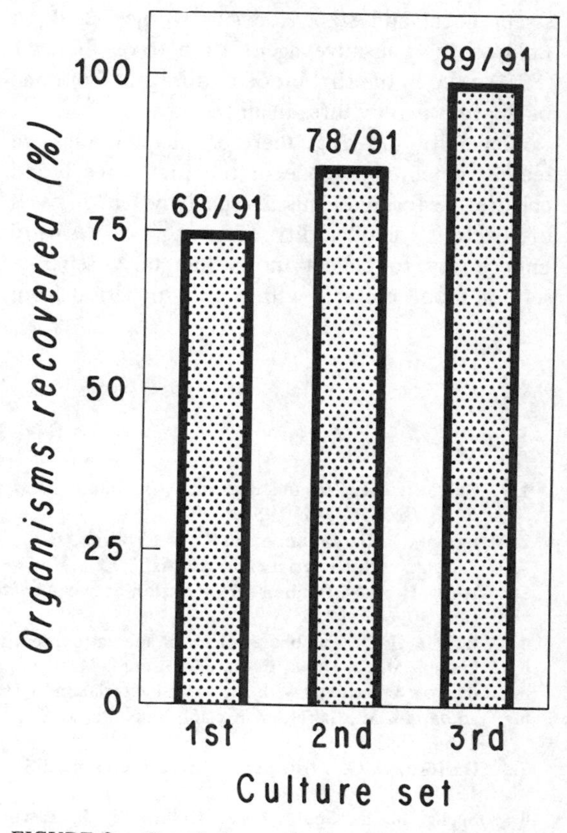

FIGURE 2. Cumulative rates of organism recovery from three blood culture sets in 80 bacteremic patients.

In a study of 80 bacteremic patients at the Mayo Clinic, Rochester, Minn., exclusive of those with endocarditis, from whom at least four separate sets of blood cultures were collected within a 14-hr period, 64 (80%) of the cultures were positive in the first set; 71 (89%) were positive in the first two sets; and 79 (99%) were positive in the first three sets (Figures 1 and 2).[19] Similar data were reported by Crowley,[20] who found that the first two blood cultures yielded the greatest number of positive results. Only 8% of cultures were positive in the second set alone, and an additional 3% were detected in the third set alone. Bartlett[21] studied 59 cases of septicemia and found 14 (24%) in which the first set of cultures was negative, 7 (12%) in which the first two sets were negative, and only 2 (3%) in which the first three sets were negative. The relationship between negativity of cultures and administration of antimicrobial agents was not determined in any of these studies.

In bacterial endocarditis, Belli and Waisbren[22] found that 52 of the 82 cases studied were diagnosed with the first culture and that in only 6 cases were more than five cultures necessary. In a similar study by Werner et al.,[11] streptococci were isolated from the first culture in 96% of cases and from one of the first two cultures in 98%. In this same study, staphylococci were isolated from the first culture in approximately 90% of instances and from one of the first two cultures in all cases of endocarditis caused by this organism. Belli and Waisbren[22] found that prior antimicrobial therapy both reduced and delayed detection of the bacteremia. Werner et al.,[11] found that antimicrobial therapy within 2 weeks prior to blood culture significantly reduced the percentage of positive cultures in 178 cases of streptococcal endocarditis, from 97 to 91% ($p < 0.02$). In cases with endocarditis due to microorganisms other than streptococci and staphylococci, Werner et al.,[11] isolated the causative agent from 9 of their 11 (82%) cases in the first blood culture and from one of the first two cultures in all cases.

Generally speaking, therefore, bacteremias are readily detected by one of the first three blood cultures performed. This is especially true in cases of bacterial endocarditis. Hence, it is usually unnecessary to collect more than three separate sets of blood cultures within a 24-hr period from patients with suspected bacterial sepsis or endocarditis, unless they have recently received or are receiving antimicrobial therapy. By the same token, it is necessary to collect more than one set of blood cultures within a 24-hr period. It would appear that the timing of collection of the culture be just prior to an anticipated rise in temperature[3] or before or at the time of a chill.[23] However, the limitations of this type of recommendation are obvious!

REFERENCES

1. **Bennett, I. L., Jr. and Beeson, P. B.**, Bacteremia: a consideration of some experimental and clinical aspects, *Yale J. Biol. Med.*, 26, 241, 1954.
2. **Beeson, P. B., Brannon, E. S., and Warren, J. V.**, Observations on the sites of removal of bacteria from the blood in patients with bacterial endocarditis, *J. Exp. Med.*, 81, 9, 1945.
3. **Weiss, H. and Ottenberg, R.**, Relation between bacteria and temperature in subacute bacterial endocarditis, *J. Infect. Dis.*, 50, 61, 1932.
4. **Wright, H. D.**, The bacteriology of subacute infective endocarditis, *J. Pathol. Bacteriol.*, 28, 541, 1925.
5. **Mallén, M. S., Hube, E. L., and Brenes, M.**, Comparative study of blood cultures made from artery, vein, and bone marrow in patients with subacute bacterial endocarditis, *Am. Heart J.*, 33, 692, 1947.
6. **Akbarian, M., Salfelder, K., and Schwarz, J.**, Experimental histoplasmic endocarditis, *Arch. Intern. Med.*, 114, 784, 1964.
7. **Harford, C. G.**, Postoperative fungal endocarditis: fungemia, embolism, and therapy, *Arch. Intern. Med.*, 134, 116, 1974.
8. **Seelig, M. S., Speth, C. P., Kozinn, P. J., Taschjian, C. L., Toni, E. F., and Goldberg, P.**, Patterns of *Candida* endocarditis following cardiac surgery: importance of early diagnosis and therapy (an analysis of 91 cases), *Prog. Cardiovasc. Dis.*, 17, 125, 1974.
9. **Rubinstein, E., Noriega, E. R., Simberkoff, M. S., Holzman, R., and Rahal, J. J.**, Fungal endocarditis: analysis of 24 cases and review of the literature, *Medicine*, 54, 331, 1975.
10. **Stone, H. H., Kolb, L. D., Currie, C. A., Geheber, C. E., and Cuzzell, J. Z.**, Candida sepsis: pathogenesis and principles of treatment, *Ann. Surg.*, 179, 697, 1974.
11. **Werner, A. S., Cobbs, C. G., Kaye, D., and Hook, E. W.**, Studies on the bacteremia of endocarditis, *JAMA*, 202, 199, 1967.
12. **Spink, W. W.**, *The Nature of Brucellosis*, University of Minnesota Press, Minneapolis, 1956.
13. **Hamilton, P. K.**, The bone marrow in brucellosis, *Am. J. Clin. Pathol.*, 24, 580, 1954.
14. **Gilman, R. H., Terminel, M., Levine, M. M., Hernandez-Mendoza, P., and Hornick, R. B.**, Relative efficacy of blood, urine, rectal swab, bone-marrow, and rose-spot cultures for recovery of *Salmonella typhi* in typhoid fever, *Lancet*, 1, 1211, 1975.
15. **DuPont, H. L. and Spink, W. W.**, Infections due to gram-negative organisms: an analysis of 860 patients with bacteremia at the University of Minnesota Medical Center, 1958–1966, *Medicine*, 48, 307, 1969.
16. **Sullivan, N. M., Sutter, V. L., Carter, W. T., Attebery, H. R., and Finegold, S. M.**, Bacteremia after genitourinary tract manipulation: bacteriological aspects and evaluation of various blood culture systems, *Appl. Microbiol.*, 23, 1101, 1972.
17. **Dietzman, D. E., Fischer, G. W., and Schoenknecht, F. D.**, Neonatal *Escherichia coli* septicemia – bacterial counts in blood., *J. Pediatr.*, 85, 128, 1974.
18. **Albers, W. H., Tyler, C. W., and Boxerbaum, B.**, Asymptomatic bacteremia in the newborn infant, *J. Pediatr.*, 69, 193, 1966.
19. **Washington, J. A., II.**, Blood cultures: principles and techniques, *Mayo Clin. Proc.*, 50, 91, 1975.
20. **Crowley, N.**, Some bacteraemias encountered in hospital practice, *J. Clin. Pathol.*, 23, 166, 1970.
21. **Bartlett, R. C.**, Contemporary blood culture practices, in *Bacteremia: Laboratory and Clinical Aspects*, Sonnenwirth, A. C., Ed., Charles C Thomas, Springfield, Ill., 1973, 15.
22. **Belli, J. and Waisbren, B. A.**, The number of blood cultures necessary to diagnose most cases of bacterial endocarditis, *Am. J. Med. Sci.*, 232, 284, 1956.
23. **Fox, H. and Forrester, J. S.**, Clinical blood cultures. An analysis of over 5000 cases, *Am. J. Clin. Pathol.*, 10, 493, 1940.

Chapter 1
SEPSIS

STATISTICAL METHODS EMPLOYED IN THE STUDY OF BLOOD CULTURE MEDIA

Duane M. Ilstrup

INTRODUCTION

When an attempt is made to evaluate the ability of various blood culture media to grow bacteria, there are three main questions that must be answered before the experiment is begun:

1. How can the experiment be designed to evaluate efficiently the culture media?
2. How can the data be collected to assure that it is as nearly free of errors as possible and that it is retrievable?
3. What statistical methods should be used when examining the data?

EXPERIMENTAL DESIGN

When studying blood culture media, there are two questions that are usually asked. The first and most important is what medium has the best chance of isolating an organism present in a patient's blood. The second is which medium will grow the organism most rapidly if two or more media have the same chance of isolating an organism.

Both of these questions can be efficiently answered in an experimental design in which the blood sample from a patient with suspected bacterial infection is inoculated simultaneously into all of the media being studied. This matched or parallel design has obvious advantages over a sequential design where either two or more media during different time intervals are evaluated, or where part of the patient population is tested with one medium and the remaining part tested with another. In a matched design, each blood sample is its own control. Each medium receives an identical sample of blood under identical conditions. Differences in patient populations, types of infections, and severity of the infection need not be of the concern they are in sequential studies.

Sample size is extremely important in these studies. Since some organisms are rarely found in a clinical environment, it is necessary to gather a sufficiently large sample of specimens to detect any differences in culture media. The general rule, to be discussed in Chapter 2, of collecting at least 2000 sets of cultures to obtain an expected 200 positive cultures seems very reasonable.

DATA COLLECTION AND RECORDS OF THE MAYO CLINIC

A variety of forms are used for data collection and record keeping. A work card (Figure 1) is used

	DATE POSITIVE	GRAM STAIN	DATE TO ANAER.	DATE OF CULTURE:
CBA				
TSB				DATE NOTIFIED:
THIOL				

SUB DATE	SOURCE	TEST	SENS. NO. & DATE	NO. & NAME OF ORGANISM	DAYS INCUB.

FIGURE 1. Blood culture work card.

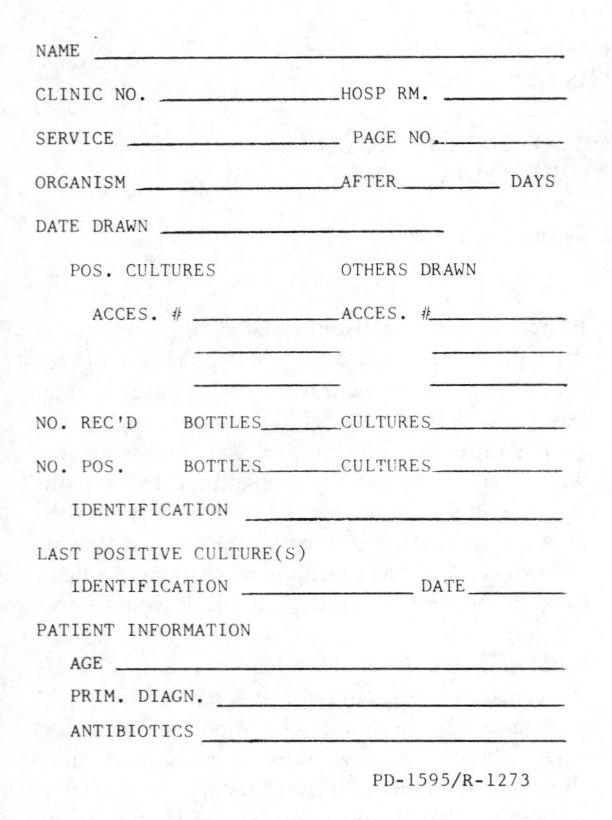

NAME ______

CLINIC NO. ______ HOSP RM. ______

SERVICE ______ PAGE NO. ______

ORGANISM ______ AFTER ______ DAYS

DATE DRAWN ______

POS. CULTURES OTHERS DRAWN

ACCES. # ______ ACCES. # ______

NO. REC'D BOTTLES ______ CULTURES ______

NO. POS. BOTTLES ______ CULTURES ______

IDENTIFICATION ______

LAST POSITIVE CULTURE(S)

IDENTIFICATION ______ DATE ______

PATIENT INFORMATION

AGE ______

PRIM. DIAGN. ______

ANTIBIOTICS ______

PD-1595/R-1273

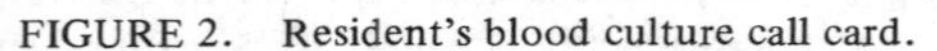

FIGURE 2. Resident's blood culture call card.

to document the initial findings in smears of blood, culture media, or of their subcultures and to follow the tests and reactions of work in progress on a positive culture. The other side of this card is marked in the appropriate spaces for keypunch entry into the laboratory's information system. A call card (Figure 2) is completed by the technologist for use by the resident when notifying a clinical service about preliminary results of cultures. Two data sheets are filled in upon completion of all work concerned with a positive blood culture. One (Figure 3) is used for recording the date of collection and accession number of the culture, the patient's location, organisms isolated and identified, and the dates of initial detection of positivity by medium. Any positive culture initially detected within the first 24 hr following its inoculation is considered to be positive in 1 day. A positive culture detected initially between 24 to 48 hr following its collection is considered to be positive in 2 days, and so on.

The data for patients who have positive cultures are abstracted from the laboratory data sheet (Figure 3) onto a computerized data collection sheet (Figure 4). This form provides space for up to four media and five organisms per specimen. Each specimen has three identification numbers: an accession number that corresponds to that on the original data collection sheet, a Mayo Clinic registration number that identifies the patient, and the date the specimen was sent to the laboratory. If an organism grows in any of the four media, its code is entered on the form and the time interval in days to its initial detection is entered for those media that were positive. A data collection form, to be implemented in the near future, will also include information about whether, initially, the detection was made macroscopically, microscopically, or by subculture, and whether sepsis or contamination was suspected.

The data from these forms are keypunched and read into a computer, and at this point are printed and scanned for errors. If any errors are found, the original records are checked and corrections are made. It must be emphasized that this is a very important part of any experiment. If reliable conclusions are to be drawn from an experiment, the experimental data must themselves be reliable.

TESTING MEDIA DIFFERENCES

Isolation Rates – Two Media

The problem of testing differences among media in the isolation rates of a single organism will be addressed first. The simplest case occurs when only two media are being investigated. The data resulting from a matched experiment on two media can be conveniently displayed in Table 1. In this table, the number of specimens that were positive for a given organism in both media is represented by the letter a, the number negative in both media by d, and the numbers positive in one medium but not in the other by b and c. The hypothesis of interest is whether the proportion of positive specimens expected with Medium 1 is the same as with Medium 2. This is tested by asking if, under the hypothesis of equal underlying proportions, the observed proportion of positives in Medium 1, $\frac{a+c}{N}$, differs from the observed proportion of positives in Medium 2, $\frac{a+b}{N}$, by more than can be reasonably predicted by chance. The easiest way to envision the appropriate test is to notice that both numerators, a + c and a + b, contain a, the number of specimens positive in both media. Hence, if the proportions are to be equal, b must equal c; i.e. the number positive in Medium 2 and

Date of Culture	Culture Number	Clinic Number	Name of Patient	Hospital or Floor	Organism	Days of Incub.	Date Pos. TSB				CBA

FIGURE 3. Log sheet for recording positive blood cultures. Columns 9 to 11 are for entry of dates of positivity of cultures in other bottles.

Data collection form for blood culture media
EVALUATION OF BLOOD CULTURE MEDIA

Column	Item	(One card per specimen)
1-6	__ __ __ __ __ __	Log book number
7-13	__ - __ __ __ - __ __ __	Clinic number
14-19	__ __ - __ __ - __ __	Date to lab (Month, Day, Year)
20	__	No. of organisms in this specimen
1st Organism		
21-23	__ __ __	1st organism
24-25	__ __	days to + medium 1
26-27	__ __	days to + medium 2
28-29	__ __	days to + medium 3
30-31	__ __	days to + medium 4
2nd Organism		
32-34	__ __ __	2nd organism
35-36	__ __	days to + medium 1
37-38	__ __	days to + medium 2
39-40	__ __	days to + medium 3
41-42	__ __	days to + medium 4
3rd Organism		
43-45	__ __ __	3rd organism
46-47	__ __	days to + medium 1
48-49	__ __	days to + medium 2
50-51	__ __	days to + medium 3
52-53	__ __	days to + medium 4
4th Organism		
54-56	__ __ __	4th organism
57-58	__ __	days to + medium 1
59-60	__ __	days to + medium 2
61-62	__ __	days to + medium 3
63-64	__ __	days to + medium 4
5th Organism		
65-67	__ __ __	5th organism
68-69	__ __	days to + medium 1
70-71	__ __	days to + medium 2
72-73	__ __	days to + medium 3
74-75	__ __	days to + medium 4
76	__	1 = If all 4 media are present, 2 = Media 1 & 2 present only, 3 = Other combination, 4 = Media 1, 2, & 3 only
78-80	4 7 1	Deck No.

FIGURE 4. Data collection form. Organism entry is by code number.

TABLE 1

		Medium 1		
		Positive	Negative	Total
Medium 2	Positive	a	b	a + b
	Negative	c	d	c + d
	Total	a + c	b + d	N = a + b + c + d

TABLE 2

		Thioglycollate		
		Positive	Negative	Total
TSB	Positive	42	27	69
	Negative	4	0	4
	Total	46	27	73

not in Medium 1 must be the same as the number positive in Medium 1 and not in Medium 2. The hypothesis of equal proportions of positivity, therefore, may be tested by asking if the numbers b and c could each have reasonably resulted by sampling from a binomial experiment of n = b + c trials with a true probability equal to one half. This is an experimental situation in which the sign test[1] is the appropriate test. The sign test is tabled[2] for experiments with n less than or equal to 90.

McNemar,[3] in 1949, derived an asymptotic chi-square test for this situation. His test, which is an approximation to the sign test, is based on the statistic $\frac{(b-c)^2}{b+c}$ which he showed to be asymptotically chi-square with one degree of freedom for sufficiently large n. Thus, if the value of this quantity is observed to be larger than the preselected percentile of this chi-squared distribution, the hypothesis will be rejected.

Table 2 is an example of the kind of an experiment that these tests are applicable to. The data are from a comparison of Trypticase soy broth (TSB) and thioglycollate – 135C. The table contains the number of positive cultures for *Pseudomonas aeruginosa* and compares the isolation rates for TSB versus Thioglycollate-135C.

The d in Table 1 is zero in this table. In general, it is not necessary to tabulate the number of specimens that were negative in both media. As noted before, it is only the number b and c that are used in the test.

Using a table of the sign test,[2] we find that the critical value for the smaller of the two numbers b or c, for a test of size $\alpha = 0.01$, is 7 for a sample size of n = b + c = 4 + 27 = 31. The smaller of the numbers b and c in this example is 4 and is less than the critical value. Therefore, it should be concluded that the proportion of positive cultures is greater for TSB than for thioglycollate.

Using McNemar's formula we have

$$\chi^2 = \frac{(27-4)^2}{27+4} = \frac{529}{31} = 17.06$$

A table of the chi-square distribution[4] shows that the p value for this test is much less than 0.01, and it can be concluded that the proportion of positive cultures is greater for TSB than for thioglycollate.

Isolation Rates – More Than Two Media

When more than two media are being compared, the data cannot be presented as conveniently as when two media are compared. The data regarding *Staphylococcus epidermidis* from an experiment comparing four media are presented in Table 3. The test is to indicate whether the proportion of positive specimens expected with Medium 1 is the same as with Media 2, 3, and 4. This example asks whether the observed proportions $\frac{41}{56}$, $\frac{30}{56}$, $\frac{17}{56}$, and $\frac{16}{56}$, differ by more than would be expected by chance.

Cochran,[5] in 1950, presented a solution to this problem that is known as Cochran's "Q"-Test. His test statistic is

$$Q = \frac{J(J-1)\left[\sum_{j=1}^{J} T_j^2 - \left(\sum_{j=1}^{J} T_j\right)^2 / J\right]}{J \sum_{i=1}^{I} u_i f_i - \sum_{i=1}^{I} u_i^2 f_i}$$

This statistic is chi-square with (J – 1) degrees of freedom where J is the number of media tested, f_1

TABLE 3

Combination i	Positive media u_i	Medium j = 1	2	3	4 = J	Specimens with combination i f_i
1	4	1	1	1	1	12
2	3	1	1	1	0	0
3	3	1	1	0	1	0
4	3	1	0	1	1	2
5	3	0	1	1	1	0
6	2	1	1	0	1	7
7	2	0	1	1	0	0
8	2	0	0	1	1	0
9	2	1	0	0	1	1
10	2	1	0	1	0	0
11	2	0	1	0	1	0
12	1	1	0	0	0	19
13	1	0	1	0	0	11
14	1	0	0	1	0	3
15	1	0	0	0	1	1
16 = I	0	0	0	0	0	0
Totals	T_j	41	30	17	16	56

is the number of specimens that were positive in all four media, f_2 is the number positive in the first three media, etc., and u is the number of media that were positive for a specimen and ranges from 0 to J or 4 in our example. T_j is the total number of specimens that were positive in Medium j. In our example we have J = 4,

$$\sum_{j=1}^{4} T_j^2 = 41^2 + 30^2 + 17^2 + 16^2 = 3126$$

$$\sum_{j=1}^{4} T_j = 41 + 30 + 17 + 16 = 104$$

$$\sum_{i=1}^{16} u_i f_i = (4 \times 12) + (3 \times 2) + \ldots + (1 \times 15) = 104$$

$$\sum_{i=1}^{16} u_i^2 f_i = (4^2 \times 12) + (3^2 \times 2) + \ldots + (1^2 \times 15) = 276$$

and

$$Q = \frac{4(3)\,[3126 - (104)^2/4]}{4\,(104) - 276} = \frac{5064}{140} = 36.1714$$

We find by looking in a chi-square table (4) with J - 1 = 3 degrees of freedom that this statistic is significant (p value < 0.001), and can conclude that the rates of positivity in the four media are not equal. If there were no significance, it would have been concluded that there was no difference in the media that could not be associated with chance. A difference was found; however, now it must be asked where the differences in the media are. The procedure, at this point, involves testing all of the four media pairwise with either McNemar's test or the sign test. Because multiple simultaneous comparisions on the same data are being done, each test must be performed at a higher level of significance to insure an overall desired level of significance. If this were done, it would be discovered that Medium 1 was not different from Medium 2 and that Medium 3 was not different from Medium 4, but that both Media 1 and 2 were significantly better than Medium 3 and Medium 4.

Time to Positivity – 2 Media

As mentioned above, a secondary question asked when testing blood culture media is, if two or more media have the same chance of isolating an organism, which medium will grow the organism most quickly. In the case of testing two media, one need only consider the time to positivity in days for the specimens that were positive in both media. Table 4 presents the experimental design.

One way to test the hypothesis that the distribution of time to positivity is the same in

TABLE 4

Time to Positivity in Days

Specimen	Medium 1	Medium 2	Medium 1 – Medium 2
1	X_{11}	X_{12}	$\Delta_1 = X_{11} - X_{12}$
2	X_{21}	X_{22}	$\Delta_2 = X_{21} - X_{22}$
3	X_{31}	X_{32}	$\Delta_3 = X_{31} - X_{32}$
⋮	⋮	⋮	⋮
a	X_{a1}	X_{a2}	$\Delta_a = X_{a1} - X_{a2}$

TABLE 5

Time to Positivity in Days

Specimen	Medium 1	Medium 2	Medium 1 – Medium 2
1	3	2	1
2	5	2	3
3	2	2	0
4	5	4	1
5	6	3	3
6	4	4	0
7	2	1	1
8	1	3	–2
9	4	1	3
10	6	2	4
Average	3.80	2.40	1.40
S^2	3.07	1.16	3.38

both media is to test whether the average time to positivity is the same in both groups. Since the pairs of readings are on the same specimen and are correlated, the two-sample t test of the equality of the means is inappropriate. The appropriate test is made on the differences between the correlated pairs and is known as the paired t test.

If the distribution of the Δs in Table 4 is approximately Gaussian, under the hypothesis of no difference in the times to positivity, the sampling distribution of the statistic

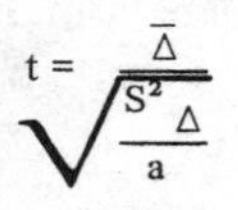

$$t = \frac{\bar{\Delta}}{\sqrt{\frac{S^2_\Delta}{a}}}$$

may be approximated by the Student's t distribution[6] having a – 1 degrees of freedom, where a is the number of specimens.

A hypothetical example of a typical two-media experiment is presented in Table 5.

The paired t test for these data is

$$t = \frac{1.40}{\sqrt{\frac{3.38}{10}}} = \frac{1.40}{0.58} = 2.41$$

By looking in a table of Student's T distribution[7] with $n - 1 = 9$ degrees of freedom, one can see that this statistic is significant (P value < 0.05) and can conclude that the average time to positivity is shorter in Medium 2 than in Medium 1. If the underlying assumption of a Gaussian distribution is not approximately met, then the paired t test may be inappropriate. A nonparametric alternative to this test is the sign test introduced earlier. In the example, the two specimens with zero differences would have to be thrown out because they are ties. An experiment of size $n = 8$ with one negative difference would remain. One finds by referring to a table of the sign test[2] that this is not quite significant ($0.05 <$ P value < 0.10).

Time to Positivity – More than 2 Media

Table 6 presents the experimental situation when testing the hypothesis of no difference in time to positivity in the case of examining more than two media.

The n specimens in this table are the specimens that were positive in all four media. The sums of the times to positivity in each specimen are represented in the table by S_1 through S_n. The sums of the times to positivity in each of the K

TABLE 6

Time to Positivity in Days

Specimen	Medium 1	Medium 2	...	Medium K	Total
i = 1	X_{11}	X_{12}	...	X_{1K}	S_1
2	X_{21}	X_{22}	...	X_{2K}	S_2
⋮	⋮	⋮	⋮	⋮	⋮
n	X_{n1}	X_{n2}	...	X_{nK}	S_n
Total	M_1	M_2	...	M_K	G

TABLE 7

Source of variation	SS	df	MS	F
Specimens	SS specimens	n–1	$\frac{\text{SS specimens}}{n-1}$	
Media	SS media	K – 1	$\frac{\text{SS media}}{K-1}$	$\frac{\text{MS media}}{\text{MS error}}$
Error	SS error	(K – 1) (n – 1)	$\frac{\text{SS error}}{(K-1)\ (n-1)}$	
Total	SS Total	Kn – 1		

media are represented by M_1 through M_K. G is the grand total of all of the data points in the table.

The experimental design presented in Table 6 is known as a repeated measures design. Winer[6] presents a derivation of the test of hypothesis for the media differences. For this test, it is assumed that the underlying distributions are again Gaussian and that the correlations between any two pair of media are equal. Whether the average time to positivity is the same in all of the media is the hypothesis being tested. The test procedures are summarized in the following analysis of variance (Table 7). In this table, SS is the sum of squares, df is the degress of freedom, and MS is the mean squared error associated with each of the sources of variation. The SS terms are computed using the following formulae:

$$\text{SS specimens} = (\Sigma S_i^2)/K - G^2/Kn$$

$$\text{SS media} = (\Sigma M_i^2)/n - G^2/Kn$$

$$\text{SS total} = \Sigma X_{ij}^2 - G^2/Kn$$

$$\text{SS error} = \text{SS total} - \text{SS specimens} - \text{SS media}$$

The test of interest is whether the medium effect is zero. This test is made by taking the ratio of the mean squared errors of media and error:

$$F = \frac{\text{SS media}/K-1}{\text{SS error}/(K-1)\ (n-1)}$$

This is an F test and has degress of freedom in the numerator of K – 1 and in the denominator of (K – 1) (n – 1), under the assumptions mentioned earlier.

The hypothetical 2-media experiment of Table 5 expanded into a 3-media experiment appears in Table 8.

TABLE 8

Time to Positivity in Days

Specimen	Medium 1	Medium 2	Medium 3	Total
1	3	2	5	10
2	5	2	6	13
3	2	2	4	8
4	5	4	4	13
5	6	3	6	15
6	4	4	6	14
7	2	1	4	7
8	1	3	2	6
9	4	1	5	10
10	6	2	5	13
Total	38	24	47	109

$G^2/Kn = 109^2/30 = 396.03333$

TABLE 9

Source of variation	SS	df	MS	F
Specimens	29.63333	9	3.2925922	
Media	26.86667	2	13.4333335	10.7626113
Error	22.46667	18	1.2481483	
Total	79.96667	29		

SS specimens $= (10^2 + 13^2 + \ldots + 13^2)/3 - 396.03333$

$= 425.66606 - 396.03333$

$= 29.63333$

SS media $= (38^2 + 24^2 + 47^2)/10 - 396.03333$

$= 422.9 - 396.03333$

$= 26.86667$

SS total $= (3^2 + 5^2 + \ldots + 5^2 + 5^2) - 396.03333$

$= 475 - 396.03333$

$= 78.96667$

SS error $= 78.9667 - 29.63333 - 26.86667$

$= 22.46667$

The analysis of variance table for this example follows:

The F statistic for differences between media is 10.76261 with 2 and 18 degrees of freedom which, according to a table of the F distribution,[8] is highly significant (P value < 0.001). It can be concluded that the average time to positivity is not the same in the three media.

If there had been no difference in the time to positivity, investigation would have stopped at this point. However, since there was a difference, it is important to find out where. One could proceed at this point as in the two media cases, testing each pair of media with the paired t test. A more conservative level of significance to insure the desired overall experimental error rate would have to be used again. The conclusions are that Medium 3 has a longer average time to positivity than either Medium 1 or Medium 2 and that Medium 1 has a longer average time to positivity than Medium 2.

If the assumption of equal pairwise correlations between media is in question, a more conservative test is appropriate. This is accomplished by assuming that the F ratio used in the previous test no longer has degrees of freedom of $K - 1$ and $(K - 1)(n - 1)$ but has degrees of freedom 1 and $n - 1$. In the examples the test would still be significant (P value < 0.01).

There are two other analytical methods that could be applied to this situation. Both will only be mentioned briefly. The first is a generalized parametric test called Hotelling's T^2 (see Winer,[6] pp. 305–308). This test should be used if the assumptions about equal correlations mentioned above are definitely not met. The second approach is a nonparametric test. It is known as Friedman's two-way analysis of variance by ranks (see Seigel,[1] pp. 166–172). In this test, no assumptions are made at all about the form of the underlying distributions.

CONCLUSIONS

The previous sections on experimental design and analytical methods were meant to introduce a microbiologist to the thinking behind the basic statistical evaluation of blood culture media. They were not meant to turn him into a statistician. If you have any questions about evaluating blood culture media and are planning to evaluate media in your own lab, please visit your friendly local statistician. He will be happy to assist you, and you won't wake up in the middle of the night wondering if you used the appropriate statistical test.

REFERENCES

1. **Siegel, S.,** *Nonparametric Statistics for the Behavioral Sciences,* McGraw-Hill, New York, 1956, 68.
2. **Beyer, W., Ed.,** *Handbook of Tables for Probability and Statistics,* Chemical Rubber Co., Cleveland, 1966, 162.
3. **McNemar, Q.,** *Psychological Statistics,* John Wiley & Sons, New York, 1949.
4. **McNemar, Q.,** *Psychological Statistics,* John Wiley & Sons, New York, 1949, 89.
5. **Cochran, W. G.,** The comparison of percentages in matched samples, *Biometrika,* 37, 256, 1950.
6. **Winer, B. J.,** *Statistical Principles in Experimental Design,* McGraw-Hill, New York, 1971, 44.
7. **Beyer, W., Ed.,** *Handbook of Tables for Probability and Statistics,* Chemical Rubber Co., Cleveland, 1966, 81.
8. **Beyer, W., Ed.,** *Handbook of Tables for Probability and Statistics,* Chemical Rubber Co., Cleveland, 1966, 96.

Chapter 2

CONVENTIONAL APPROACHES TO BLOOD CULTURE

John A. Washington II

COLLECTION

Skin Antisepsis

Despite the great variability in the numbers of bacteria recoverable from adjacent skin sites,[1] it is well known that large numbers of Gram-positive and Gram-negative bacteria and a great variety of yeasts and filamentous fungi can be recovered from the skin.[2] That these organisms are predominantly saprophytic in nature is of little help in interpreting the results of blood cultures from which they are recovered in view of their potential significance, especially in causing endocarditis and infections of implanted prosthetic material. Gram-negative bacilli are relatively uncommon on normal, healthy skin; however, these organisms are often recovered from the skin of hospitalized patients and nursing personnel who have recently attended such patients.[2] Thus, ample opportunities for contamination of blood cultures are provided by the patient's or the phlebotomist's own skin microflora.

The need for careful skin antisepsis is obvious but, unfortunately, easily forgotten and often neglected. In certain instances, however, even scrupulous care has serious limitations. Within the time limits ordinarily employed for preparation of the skin for venipuncture, Ahmad and Darrell[3] showed that a 30-sec scrub with 70% isopropyl alcohol was relatively ineffective in eliminating *Clostridium perfringens* from the skin of the antecubital fossa. They demonstrated that *C. perfringens* was present in this area in approximately 20% of 185 inpatients studied and that these organisms remained in 16% of the patients following skin preparation. Most of these patients were elderly, bedridden, and chronically hospitalized females with underlying orthopedic problems. Since 70% isopropyl alcohol is not sporicidal, these results are not particularly surprising and simply confirmed earlier studies by Ayliffe and Lowbury.[4] Moreover, Updegraff[5] showed that although a 1-min scrub with 70% ethyl alcohol reduced the skin microflora by 90%, it did not kill all of the bacteria in any of 14 layers of skin tested. Ahmad and Darrell[3] did not extend their studies by following their alcohol scrub with an iodine or iodophor scrub.

While there are ample data documenting the efficacy of iodine and iodophors in significantly reducing skin microflora, these data can be applied in only limited fashion to the usually cursory method of antisepsis used in preparation of the skin for blood culture collection. Rapid degerming activity by iodophor preparations has been well documented,[6] but these findings must be carefully interpreted within the limits and conditions of the experimental design used in the studies.

In view of this, one point is clear – that instant antisepsis does not occur, barring use of an agent that is itself instantly destructive to the skin and subcutaneous tissues. In practice, tincture of iodine and alcohol represent the agents most commonly used.[7] It is generally recommended, because of the residual effect of iodine or iodophor, that the alcohol (70% ethyl or isopropyl) be applied first to clean the skin and that it be followed by a 1 to 2% tincture of iodine or an iodophor preparation which is applied in concentric fashion.[8,9] Some patients do give a history of hypersensitivity to iodine and, therefore, object to its use. For this reason, a commercially available, prepackaged unit containing three large swabs soaked with iodophor is being used at the Mayo Clinic. One swab is used to scrub the venipuncture site, the second to prepare it concentrically, and the third to apply to the rubber stoppers of the blood culture bottles. This technique kept the incidence of contamination in blood cultures collected from 240 adults without infection at an acceptably low level of 2.1%.[10] In a survey of 21 laboratories by Bartlett,[7] 13 (62%) reported that

5% or more of their blood cultures were probably contaminated. At the Mayo Clinic, monthly contamination rates vary between 2 and 3%[12] (Figure 1).

The hazards of using aqueous benzalkonium chloride solution for skin antisepsis were emphasized in a report of nosocomial pseudobacteremia by Kaslow et al.,[11] who investigated a community hospital in which the reported incidence of bacteremias due to *Pseudomonas cepacia* and *Enterobacter* had increased substantially during the previous year. Storage bottles of this antiseptic and saturated cotton swabs on venipuncture trays were found to contain combinations of *P. cepacia, Enterobacter agglomerans, Enterobacter cloacae,* and *Serratia marcescens.* Replacement of benzalkonium chloride with iodine and alcohol for skin antisepsis virtually eliminated the problem overnight. Despite the fact that benzalkonium chloride has been reported often to have been a source of contamination in hospitals and to have failed to disinfect consistently, it continues to be used for skin antisepsis, while it seems evident it should not be used for purposes of disinfection.

It is unfortunate that the collection of blood for culture is generally performed by physicians[7] who are usually busy, frequently harassed, and often unaware of the need for or unable to devote the time to careful aseptic technique. Much time and attention are given by the laboratory to the examination of blood cultures and to the isolation, identification, and antimicrobial susceptibility testing of any organisms present. In most instances, the results of blood cultures are carefully considered by the medical staff attending the patient, and, in many instances, these results serve to commit a patient to a prolonged and expensive course of antimicrobial therapy. While the author does not intend to deprecate the physician's ability to assess the results of this entire process, it is puzzling why daily trials and tribulations should be compounded by adhering to a faulty procedure. In this cost-conscious era, no one has estimated the vast sums of money that are expended on

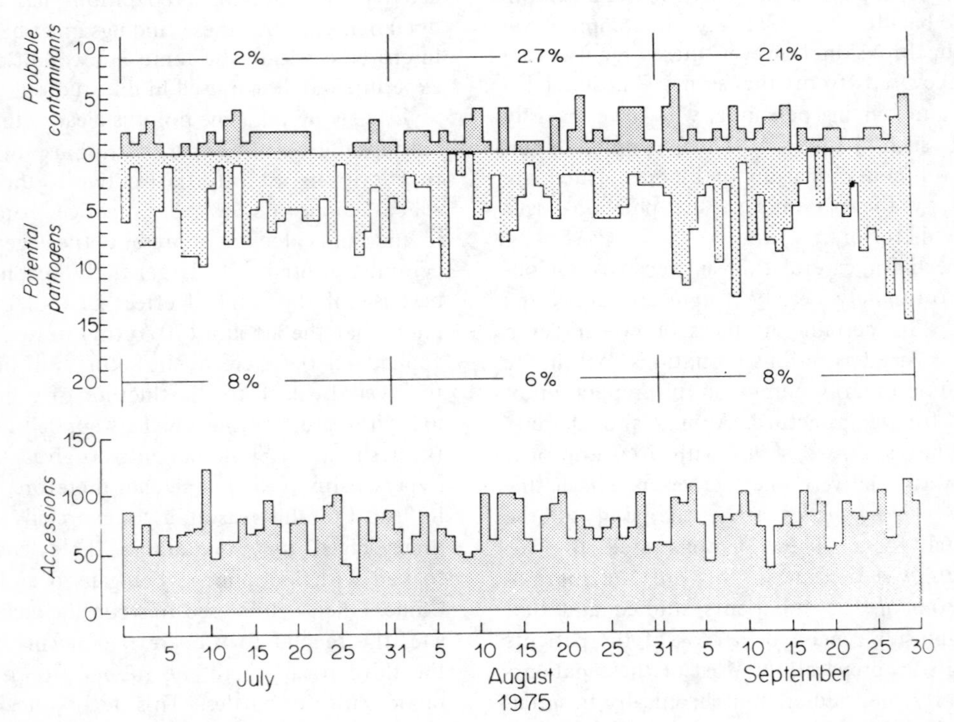

FIGURE 1. Quality control record during 3 months of 1975 illustrating number of blood-culture sets accessioned daily and number of probable contaminants and potential pathogens isolated daily. (From Washington, J. A., II, *Significance of Medical Microbiology in the Care of Patients,* p. 43. © 1977 The Williams & Wilkins Co., Baltimore. With permission.)

contaminated cultures. It would appear that less time and fewer dollars would be expended by developing teams of laboratory personnel who are trained in venipuncture techniques and who can, along with their other duties, properly collect blood cultures. A number of years ago, at the Clinical Center, National Institutes of Health, Bethesda, Md., a comparison was made between the contamination rates in blood cultures collected by technologists during the day with those in blood cultures collected by house officers at night. The latter rates were more than double the former. Experience and aseptic technique undoubtedly were important variables which contributed to this difference. An analogous situation relating experience to colonization rates of subclavian vein catheters was reported by Bernard et al.[13]

Without belaboring this issue further, the availability of a venipuncture team should reduce contamination rates to between 2 and 3% of all blood cultures collected. In neonates, the problems associated with blood collection are greater, as is the likelihood of contamination. Dunham,[14] Waddell et al.,[15] Eitzman and Smith,[16] and Albers et al.[17] have reported contamination rates ranging from 14 to 25% despite seemingly adequate skin antisepsis. This problem is compounded when the blood is taken from the umbilical vein, and contamination rates of up to 45% have been reported in such instances.[18,19] Balagtas et al.[20] found that 32% of newborn infants with positive umbilical cord cultures had positive cultures of blood collected from umbilical vein catheters. Whether blood is collected from the umbilical vein or from a peripheral vein appears to influence significantly the outcome of cultures from newborn infants, and, in all likelihood, this accounts for the major discrepancies in reported incidences of anaerobic bacteremias in newborn infants. Chow et al.[21] reported that 26% of neonatal bacteremias were associated with anaerobes, while only 2.9% were found by Thirumoorthi et al.[22] to have been caused by anaerobes. In the former study, cord blood specimens accounted for slightly more than one third of the anaerobic bacteremias; only peripheral venous blood was cultured in the latter study. One possible solution to this problem was suggested by Cowett et al.,[23] who reported that only 1.8% of 113 blood cultures collected by umbilical artery catheterization were contaminated. As a general recommendation, Gotoff and Behrman[24] suggested that unless contraindicated by the patient's condition or size, two blood samples from peripheral veins should be collected for culture.

Obviously, there are a number of other sources of contamination in blood culture, which will be reviewed below; however, the importance of skin as one source cannot be overlooked. Despite its shortcomings, careful skin antisepsis is an important step in the process of blood culture.

Timing and Number of Cultures

As discussed in Chapter 1 (Washington), it would appear that the optimal time for the collection of blood for cultures is just prior to the anticipated onset of a chill. Obviously, few of us are able to predict with any degree of certainty when this event will occur. Typically, therefore, blood is collected at the time of a rise in temperature or the onset of chills. From what has already been stated, however, it is imperative that more than one set of cultures be inoculated and that the blood collections be made separately. The time interval between blood collections can be arbitrarily set at 1 hr; however, in medically urgent situations in which antimicrobial therapy is to be started immediately, it is advisable to make two separate blood collections within a few minutes' time. A general recommendation that these collections be made separately from each arm may be helpful, since it is often tempting for the phlebotomist to collect a large volume of blood and then to distribute it among two or more blood culture bottle sets. The obvious risk posed by this practice is that of contamination of this large volume of blood and the subsequent growth of the contaminant in multiple bottles.

Ideally, three separate sets of blood cultures should be inoculated with blood collected by three separate venipunctures within a 24-hr period, and, preferably, within a 12-hr period after the onset of symptoms.[25] The rationale for this recommendation has already been thoroughly discussed (Chapter 1). In certain instances, such as the occurrence of another fever spike or chill, it may be advisable to collect blood for additional cultures. Certain policies have been implemented at the Mayo Clinic regarding this issue which have been very helpful. A physician's order for a single blood culture is routinely duplicated so that two separate sets are collected. Orders for more than three sets of blood cultures within a 24-hr period must be approved by a resident in clinical microbiology who, in turn, is responsible for notifying the venipuncture team of his approval for it to collect the additional cultures. Although some requests for more than three cultures per 24 hr are legitimate, many are due to the fact that a consulting physician is often unaware of the number of cultures already ordered and in process.

In cases with bacterial endocarditis the timing between blood collections is not critical, since the bacteremia is continuous. It would seem reasonable to recommend the collection of three separate sets of blood cultures at hourly intervals on each of two consecutive days and then to await the results of these cultures before collecting additional blood.

The issue of culturally negative cases of bacterial endocarditis remains unresolved. Various studies, including those by Belli and Waisbren,[26] Blount,[27] Cherubin and Neu,[28] and Hayward,[29] have cited percentages of culturally negative cases of bacterial endocarditis in the range of 17 to 41%. To some extent, the variability of these figures reflects differences in the criteria used to make a clinical or post-mortem diagnosis of endocarditis; however, Werner et al.[30] did find in 206 cases of endocarditis that 95% of blood cultures were positive, and in 91% of cases all blood cultures were positive. Similar results were reported by Barritt and Gillespie[31] and by Cates and Christie.[32] The reasons for these discrepancies remain speculative and could include prior antimicrobial therapy, volume of blood cultured, media employed, and, as already stated, the criteria used for the diagnosis of endocarditis in the absence of positive ante-mortem cultures.

The unreliability of post-mortem heart blood cultures in helping to establish the diagnosis of bacterial endocarditis was demonstrated many years ago by, among others, Wright,[33] who in 1925 found 37% of 100 such specimens cultured from cases without endocarditis to be positive. In most of these cases (22%), streptococci were isolated in pure or mixed culture. In a more recent study of post-mortem cultures of heart blood from cases with no apparent infection, Wilson et al.[34] found 7% to be positive.

Volume

The volume of blood collected for culture is important for two reasons: first and foremost, it bears directly on the laboratory's ability to detect bacteria in blood cultures, and second, it is a major, but often overlooked, variable in studies comparing the efficacy of blood culture media or systems. A major point of confusion is the relationship between the volume of blood collected for culture and the number of cultures recommended. Putting this issue another way, can the number of cultures performed be reduced as the volume of blood collected is increased? The answer, simply stated, is no. The number of cultures performed and the volume of blood collected are two questions which must be addressed separately in the laboratory. Obviously, if a patient has 20 sets of blood cultures performed over a 48-hr period, the volume of blood lost becomes a real problem; however, the laboratory which permits this practice to go on might be considered negligent. A number of years ago, the Mayo Clinic had a patient with subacute bacterial endocarditis from whom 19 sets of blood cultures were received within 48 hr, all of which contained viridans streptococci. Clearly, there was no justifiable reason, medically or otherwise, for this excessive number of specimens, and a policy was subsequently implemented to prevent its recurrence. The question about the number of blood cultures needed to isolate the etiologic agent of bacteremia has already been addressed (Chapter 1). Two to three separate cultures will yield the etiologic agent in at least 90% of instances.

Varying amounts of blood have been collected for culture from newborn infants. Eitzman and Smith[16] withdrew 10 ml of blood from the internal jugular vein; 4 ml were generally collected by Albers et al.[17] Minkus and Moffet[35] cultured 1 to 5 ml of blood, and Franciosi and Favara[36] cultured 1 to 3 ml. Smaller volumes of 0.5 to 1.0 ml were used by Cowett et al.[23] By documenting the order of magnitude of bacteremia in newborn infants, Dietzman et al.[37] were able to demonstrate that in the majority (77%) of *Escherichia coli* septicemias, culturing as little as 0.2 to 0.5 ml of blood would have been expected to yield positive results; however, they questioned whether cultures of less than 1 ml of blood would have been adequate in the remainder of cases in which there were colony counts of fewer than 5 per milliliter. As a general rule, therefore, it would seem reasonable to recommend a collection of 1 to 3 ml of blood from newborn infants.

Few studies have been performed in which the rates of recovery of bacteria from parallel cultures of different volumes of blood were the only variables examined. In one such study reported by Hall et al.,[38] parallel cultures of 5 and 10 ml of blood were performed in transiently vented 100-ml bottles containing Tryptic soy broth with sodium polyanetholsulfonate (SPS) and CO_2. The cultures of the larger volume of blood yielded significantly more Gram-negative bacilli than did those of the smaller volume ($p < 0.01$); no significant differences between cultures of the two volumes were noted with Gram-positive cocci or yeasts (Table 1). Comparing cultures of 2 and 5 ml of blood in supplemented peptone broth with SPS, Tenney et al.[39] found that significantly more organisms, and especially Gram-negative bacilli, were isolated from cultures of the larger volume than from the smaller volume of blood ($p < 0.05$). These investigators found an average increase in detection of bacteria of 4.5% per milliliter of blood cultured. Tenney et al.[39] concluded that when evaluating other factors affecting the detection of bacteremia, comparisons are valid only when the volumes of blood cultured are equal. To expand this theme, it is critical for the readers of articles reporting evaluations of blood culture systems or media to study the fine print in the section describing the materials and methods used in the study. There are numerous variables affecting the results of blood culture, analysis of any one of which in a comparative study must be questioned if the volumes of blood cultured are not comparable.

Obviously, the results of Hall et al.[38] and those reported by Tenney et al.[39] pose a serious dilemma for the laboratory, in that it has been clearly demonstrated that the more blood cultured the better; however, neither study has established what the optimal volume of blood to be cultured is. In analyzing the data of Hall et al.[38] in a manner similar to that used by Tenney et al.,[39] an increase in volume of blood cultured from 5 to 10 ml increased the overall recovery of organisms, excluding presumed contaminants, by nearly 15% or nearly 3% per additional milliliter of blood cultured (Figure 2). By plotting the combined rate of recovery of all organisms but the presumed contaminants from 5 and 10 ml of blood, the yield, relative to 5 ml alone, increases in linear

TABLE 1

Numbers of Isolates[a] in Blood Culture Bottles Containing 100 and 50 ml of TSB

	Isolates (no.) in			
Isolate	100-ml bottle only	50-ml bottle only	Both 100- and 50-ml bottles	p value
Gram-negative bacilli				
Enterobacteriaceae	23	10	74	
Pseudomonadaceae	13	1	31	
Bacteroidaceae	3	3	12	
Other	1	0	0	
Subtotal	40	14	117	<0.01
Gram-positive cocci				
Micrococcaceae	15	9	42	
Streptococcaceae	7	6	48	
Peptococcaceae	0	0	1	
Subtotal	22	15	91	NS[b]
Yeasts				
Candida	2	2	8	NS

[a]Exclusive of single sets of cultures positive for *Bacillus, Corynebacterium, Propionibacterium,* and *Staphylococcus epidermidis* (presumed contaminants).
[b]NS, not significant.

From Hall, M. M., Ilstrup, D. M., and Washington, J. A., II, *J. Clin. Microbiol.*, 3, 643, 1976. With permission.

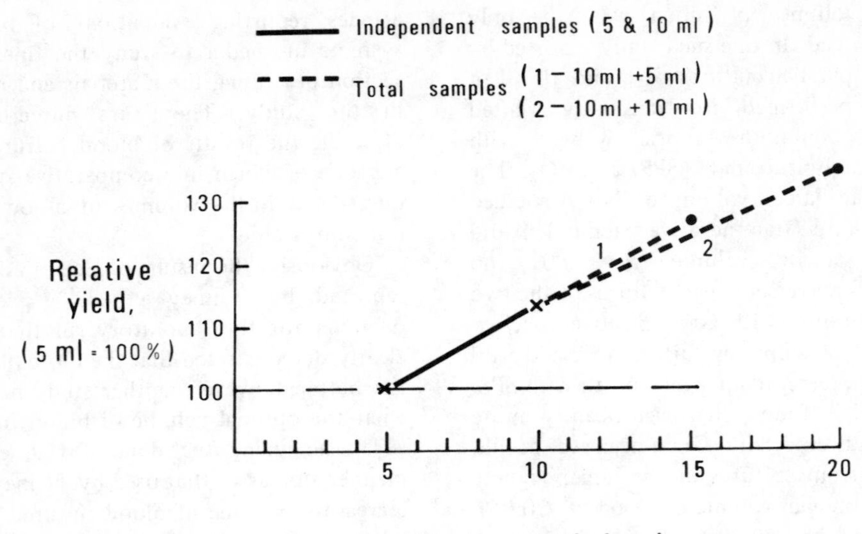

FIGURE 2. Recovery of organisms by volume of blood cultured.

fashion (Figure 2). In a separate study, the combined rate of recovery of all organisms, except for presumed contaminants, from cultures of duplicate 10-ml samples increased in slightly less than linear fashion (Figure 2). Since 5 ml has been found to yield significantly more isolates than 2 ml,[39] since 10 ml has yielded significantly more isolates than 5 ml,[38] and since 20 ml has yielded more isolates than 10 or 15 ml, it is recommended that a larger volume be used for culture. At the Clinic, a routine collection is made of 20 to 30 ml from adults, dividing 20 ml equally between two 100-ml blood culture bottles, one of which is vented transiently prior to incubation and the other of which remains unvented. The residual 10 ml, when collected, is inoculated into a third 100-ml bottle which is also vented transiently.

Again, the most important point about both of these studies is that the volumes of blood cultured were the only variables examined. The media, additives, atmospheres and duration of incubation, and subculture techniques were held constant. The lack of control of these other variables largely invalidates any firm conclusions regarding the influence of volume of blood cultured on the rate of positivity of cultures in other studies.

Specimen Handling

Blood may be collected for culture with a sterile needle and syringe and then inoculated directly into blood culture bottles. Alternatively, a bleeding or transfer set may be used. There are no data documenting the relative rates of contamination of blood cultures using these two methods. The former method is employed at the Mayo Clinic simply because blood is often to be collected for tests other than for cultures.

In some hospitals, an evacuated blood collection tube containing SPS is used to transport the blood from the patient to the laboratory where it is inoculated into blood culture media. This system was described by Ellner in 1968[40] and was found to reduce the contamination from the 8.2% level observed in open flasks to 2.4% in one study of 4426 cultures and to 1.4% in a later study of 8382 cultures. Unfortunately, these studies were not parallel, and no well-controlled parallel studies have been performed comparing direct inoculation of media with inoculation of media following the transport of blood in a tube containing SPS. The advantages of direct inoculation of blood culture media at the bedside relate primarily to the earlier initiation of the growth phase of any organisms which are present, thereby providing the opportunity for the earlier detection of growth.

In recent years, one manufacturer (Becton, Dickinson and Company, Rutherford, N.J.) has introduced a blood culture system in which supplemented peptone broth with SPS was included in an evacuated blood collection tube (Vacutainer®). The problems associated with the small volume of blood cultured in this system have already been discussed and have been documented by Tenney et al.[39] The company has subsequently manufactured Vacutainer blood culture bottles containing 45 ml of media. Although the direct inoculation of blood into media in this system seems attractive and convenient, it is important to introduce a word of caution about this approach. A backflow hazard from evacuated blood collection tubes has been documented by Katz et al.,[41] who determined that a series of bacteremias with *Serratia marcescens* were caused by backflow from nonsterile tubes. Hoffman et al.[42] traced a substantial increase in bacteremias due to *S. marcescens* to the transfer of organisms from contaminated pediatric sized vacuum tubes containing ethylenediamine tetraacetic acid (EDTA) to blood culture media. Crosscontamination resulted when physicians simultaneously obtained blood for counts and cultures, injected blood into the EDTA-containing vacuum tube first, and then changed needles to inoculate the blood culture bottle. Backflow from the EDTA-containing tube occurred when pressure on the plunger of the syringe was released, thereby contaminating the residual blood in the syringe and leading to spuriously positive blood cultures. Obviously, this hazard is not limited to the backflow of microorganisms which may be present in nonsterile tubes, but includes the backflow of additives, including media, from such tubes. Moreover, a multicenter, collaborative study of the microbiology of evacuated blood collection tubes has suggested that the hazards of backflow of microorganisms may not be limited to nonsterile tubes.[43] At any rate, the use of such blood collection systems requires extreme care, including holding the tube or bottle upright so that the proximal end of the needle does not become immersed in the medium, removing the tourniquet well before the blood flow into the tube or bottle has stopped, and avoiding movement of or pressure on the end of the tube or bottle.[41]

BACTERIAL BLOOD CULTURES

Media

The variety of media which has been described for use in blood cultures is almost infinite, particularly when one considers the types of additives which may be included in these media. Soybean-casein digest, Columbia, brain heart infusion, thiol, thioglycollate, Brucella, supplemented peptone, and dextrose phosphate broths are those most frequently available commercially and probably remain those most commonly used.[7] It is remarkable, however, that there have been relatively few parallel studies comparing the performance of these media in clinical use. Such parallel studies are essential so that the selection of media for routine use can be made rationally. Sequential studies comparing the isolation rates of bacteria in two or more different media during different time intervals simply lack validity. There are changes in patient populations, rates and types of nosocomial infections, and physicians' ordering habits, to mention only a few variables which preclude comparison of the performance of Medium X during the first 6 months of 1 year, for example, with that of Medium Y during the second 6 months of the same year.

Comparisons of media or blood culture systems, therefore, need to be made in parallel. Moreover, they require that a large number of cultures be collected under well-controlled conditions. As a general rule, it is necessary to collect at least 2000 sets of cultures in order to be able to compare isolation rates of various groups of bacteria with any degree of confidence. Even then, there will obviously be groups of infrequently or rarely encountered bacteria which cannot be examined statistically. Most laboratories find that fewer than 14% of their cultures become positive,[7] and the Clinic's experience has been that 8 to 10% of all of the cultures become positive, including those with presumed contaminants. Therefore, one might plan in a blood culture study of 2000 sets of cultures on having 200 positive sets of cultures for statistical analysis.

Obviously, one is limited in such studies by the volume of blood which can reasonably be collected; however, it is essential that the volumes of blood inoculated into the media under study be equal. For this reason, it is difficult to conduct such studies without the participation and cooperation of a venipuncture team.

Composition

Over the years, the Clinic has conducted a number of studies comparing the performance of various blood culture media. From the beginning, soybean-casein digest broth has served as the basis for comparison with other media. In most instances, tryptic soy broth, or TSB (Difco

Laboratories, Detroit, Michigan) has represented the soybean-casein digest broth product used, primarily for reasons of quality, cost, supply, and service; therefore, the use of this product should not be misconstrued as an implied endorsement of it. The Clinic has never compared different manufacturers' soybean-casein digest broths and can, therefore, make no comments about their relative effectiveness. Such a comparative study would, of course, be of interest; however, prior to initiating such studies, it would probably be necessary to establish what the reproducibility of results in the "reference" broth would be. In other words, it would be necessary to test one particular lot of TSB in duplicate under identical conditions of inoculation, incubation, examination, and subculture. Recently, such a comparison was made at the Clinic with two transiently vented 100-ml bottles containing TSB with 0.025% SPS without, however, ensuring that all bottles came from the same production lot. Of 233 positive cultures, 56% were positive in both bottles, 25% in only one bottle, and 19% in only the other bottle. This kind of distribution of results is not surprising in view of the important role played by the volume of blood cultured and the likelihood that those few bacteria present in blood are not evenly dispersed.

Even if one were to assume identity of composition of different manufacturers' media, e.g., soybean-casein digest broth, it is highly unlikely that their production and bottling processes are identical. The degrees of vacuum and the amounts of CO_2 and dissolved O_2 existing in bottles, for example, quite probably vary from manufacturer to manufacturer and could even vary from lot to lot. Over the years, vast differences have been found in the performance of the same basal medium bottled by different manufacturers. In one such instance, for example, despite the fact that the concentrations of SPS were stated to be equal, a 10% v/v addition of blood consistently clotted in one manufacturer's bottle and not in the other's. Moreover, there was good reason to believe that the atmospheric conditions in these two manufacturers' unvented bottles differed substantially, since yeasts were consistently isolated more frequently and anaerobes less frequently from one than from the other.

With the exception of 1968 and the first quarter of 1969 in which two 50-ml bottles were used, blood from patients at the Mayo Clinic has routinely been inoculated into at least two, and usually three, 100-ml bottles on a 10% v/v basis. It was not until 1976, however, that the theoretical reason for having decided in 1969 to culture this larger volume of blood was given factual substance.[38]

Thiol (Difco Laboratories, Detroit, Michigan) and TSB were compared in the first study of 3795 positive cultures at the Mayo Clinic between 1968 and 1970.[44] Both bottles contained 100 ml of media under vacuum with CO_2; neither contained SPS nor was vented. Both were examined daily for 14 days and were subcultured routinely if macroscopically negative after 48 hr of incubation. The number of isolates by medium are listed in Table 2. Statistically significant differences in the rates of recovery of various bacteria and yeasts were limited to *Actinobacillus* and *Pseudomonas,* which were recovered more frequently from TSB than from thiol ($p < 0.05$ and $p < 0.001$, respectively) and to corynebacteria and streptococci, which were recovered from Thiol more frequently than from TSB ($p < 0.001$ and $p < 0.005$, respectively). In those instances in which both media were positive, the time intervals to detection of positivity were significantly shorter ($p < 0.05$) in TSB for *Actinobacillus, Enterobacter,* and *Pseudomonas,* and in thiol for *Proteus* and *Streptococcus;* however, only in the cases of *Actinobacillus* and *Proteus* did these differences exceed 1 day. In this study, anaerobic bacteria represented 11% of the positive cultures and 20% of the patients, but there were no statistically significant differences in their rates of recovery or in the time intervals required for their detection in these media, despite the fact that the redox potential (E_h) in thiol is substantially lower than that in TSB.[135] Thiol is described by its manufacturer as possessing the ability to neutralize the bacteriostatic and bactericidal activities of streptomycin, penicillins, and sulfonamides; however, an examination of the cultures of 70 patients with streptococcal endocarditis failed to show any significant differences between the recovery rates of streptococci from thiol and TSB, irrespective of the administration of antimicrobial agents within 2 weeks prior to blood cultures. These data are at variance with those reported by Werner et al.,[30] who did find significantly fewer positive cultures in patients who had received antimicrobial agents within 2 weeks prior to their blood cultures. On this basis, one might well have anticipated a significant difference in recovery

TABLE 2

Number of Isolates in Positive Cultures by Media

Organism	No. positive with TSB[a] and Thiol	TSB	Thiol	Total positive	p (χ^2 analysis)
Actinobacillus	14	7	1	22	<0.05
Bifidobacterium eriksonii	2	0	3	5	NS[b]
Clostridium	22	8	14	44	NS
Bacillus	16	65	82	163	NS
Corynebacterium	66	126	203	395	<0.001
Escherichia coli	302	120	111	533	NS
Shigella	0	0	1	1	NS
Salmonella	6	4	3	13	NS
Citrobacter	6	0	0	6	NS
Klebsiella	140	54	51	245	NS
Enterobacter	55	17	18	90	NS
Serratia marcescens	37	8	9	54	NS
Proteus	43	23	22	88	NS
Providencia	2	0	0	2	NS
Haemophilus	12	8	11	31	NS
Diplococcus pneumoniae	35	16	9	60	NS
Listeria monocytogenes	6	4	4	14	NS
Streptococcus	519	99	148	766	<0.005
Eubacterium lentum	1	0	2	3	NS
Herellea vaginicola	4	7	3	14	NS
Alcaligenes faecalis	6	7	4	17	NS
Mima polymorpha	0	2	0	2	NS
Pasteurella multocida	1	0	0	1	NS
Neisseria	0	5	0	5	NS
Bacteroidaceae	182	74	76	332	NS
Micrococcus	2	1	2	5	NS
Staphylococcus aureus	258	83	93	434	NS
Staphylococcus epidermidis	80	105	116	301	NS
Peptostreptococcus	13	5	9	27	NS
Peptococcus	4	2	7	13	NS
Veillonella	0	1	0	1	NS
Pseudomonas	57	131	25	213	<0.001
Aeromonas	0	1	1	2	NS
Candida	9	9	6	24	NS
Torulopsis glabrata	0	0	4	4	NS

[a]TSB, Tryptic soy broth.
[b]NS, not significant.

From Washington, J. A., II, *Appl. Microbiol.,* 22, 604, 1971. With permission.

rates between Thiol and TSB in the patients who had recently received antimicrobial agents, although, for unclear reasons, this was not the case.

A second study of 1116 positive cultures at the Mayo Clinic compared TSB with thioglycollate-135C (Becton-Dickinson, Rutherford, N.J.).[4 5] Both bottles contained 100 ml of media under vacuum with CO_2; neither was vented nor contained SPS. Subcultures were routinely made of macroscopically negative bottles after 24 hr of incubation, and all bottles were examined daily for 14 days. As may be seen in Table 3, the statistically significant differences in recovery rates of bacteria and yeasts were limited to *Corynebacterium* and *Staphylococcus epidermidis,* which were recovered more frequently from thioglycollate-135C than from TSB ($p < 0.01$ and $p < 0.05$, respectively), and to *Pseudomonas aeruginosa,* which was recovered more frequently from TSB than thioglycollate-135C ($p < 0.01$). There were no significant differences between these media in the time intervals required for detection of growth. Anaerobic bacteria were not recovered significantly more frequently from the thioglycollate broth.

In 1973, blood culture media containing SPS were introduced into routine use at the Mayo Clinic, and a comparison between TSB and thiol, both containing this additive, was made with 611

TABLE 3

Number of Isolates in Positive Cultures, by Media[a]

Organism	No. positive				
	TSB and thio	TSB	Thio	Total positive	p (χ^2 analysis)
Bacillus	1	10	11	22	NS[b]
Bifidobacterium	0	0	1	1	NS
Clostridium	3	0	4	7	NS
Corynebacterium	21	33	103	157	<0.01
Escherichia coli	123	59	43	225	NS
Salmonella	0	1	2	3	NS
Arizona	2	1	3	6	NS
Citrobacter	1	0	0	1	NS
Klebsiella	23	5	9	37	NS
Enterobacter	7	4	6	17	NS
Serratia	17	2	3	22	NS
Proteus	20	7	5	32	NS
Haemophilus	5	4	5	14	NS
Diplococcus pneumoniae	19	4	6	29	NS
Streptococcus, unspecified	1	0	1	2	NS
Streptococcus, viridans group	82	12	18	112	NS
Streptococcus, group A	11	3	1	15	NS
Streptococcus, group C	6	0	0	6	NS
Streptococcus, group D	25	10	8	43	NS
Streptococcus, other groups	3	0	0	3	NS
Eubacterium	0	2	2	4	NS
Herellea vaginicola	6	8	2	16	NS
Alcaligenes faecalis	4	3	3	10	NS
Mima polymorpha	0	0	1	1	NS
Neisseria	3	1	0	4	NS
Bacteroidaceae	54	20	18	92	NS
Micrococcus	1	1	1	3	NS
Staphylococcus aureus	60	19	32	111	NS
Staphylococcus epidermidis	28	22	40	90	<0.05
Peptostreptococcus	9	0	0	9	NS
Peptococcus	7	1	4	12	NS
Pseudomonas aeruginosa	42	27	4	73	<0.01
Candida	12	6	8	26	NS
Torulopsis glabrata	1	0	2	3	NS

[a]TSB = trypticase soy broth; thio = thioglycollate.
[b]NS, not significant.

From Washington, J. A., II, *Appl. Microbiol.,* 23, 956, 1972. With permission.

positive cultures.[46] Both bottles remained unvented and were routinely subcultured after 24 hr of incubation. The statistically significant differences in rates of recovery of bacteria were restricted to corynebacteria and *Pseudomonas,* which were recovered more frequently ($p < 0.001$) from TSB than from Thiol (Table 4). *Escherichia coli, Haemophilus,* and the Bacteroidaceae were also recovered more frequently from TSB than from Thiol; however, these differences were not statistically significant. The mean times required to detect growth also did not differ significantly.

During this study, blood was inoculated in parallel into two tubes of supplemented peptone broth (Vacutainer Culture Tubes), as well as into TSB and Thiol. The bottles with TSB and Thiol each contained 100 ml of media, and neither of these bottles was vented. The tubes each contained 18 ml of broth and were each inoculated with 2 ml of blood; only one tube was vented. As seen in Table 5, which lists only those groups of organisms in which differences in rates of detection occurred, the culture tubes failed to detect a large number and variety of potentially pathogenic bacteria and yeasts in comparison with TSB and Thiol. In a comparison of the culture tube with thioglycollate broth and with hypertonic Brucella and Trypticase

TABLE 4

Isolates in Positive Cultures, by Medium

Organism	No. positive in					
	TSB and Thiol	TSB only	Thiol only	Total positive[a]	p[b]	Adjusted percent positive[c]
Bacillus	0	7	7	14	NS	
Clostridium	2	1	1	4	NS	0.8
Corynebacterium	8	70	20	98	<0.001	
Escherichia	87	22	10	119	<0.1	25.1
Salmonella	2	0	0	2	NS	0.4
Citrobacter	2	2	1	5	NS	1.1
Klebsiella	35	12	5	52	NS	10.9
Enterobacter	6	3	2	11	NS	2.3
Proteus	11	3	2	16	NS	3.4
Haemophilus	5	5	0	10	<0.1	2.1
Streptococci						
S. pneumoniae	14	3	3	20	NS	4.2
Group A	6	1	1	8	NS	1.7
Group D	27	5	4	36	NS	7.6
Other groups	0	0	1	1	NS	0.2
Viridans	30	4	5	39	NS	8.2
Acinetobacter	0	4	1	5	NS	1.1
Alcaligenes	1	1	0	2	NS	0.4
Bacteroidaceae	23	10	3	36	<0.1	7.6
Staphylococci						
S. aureus	45	20	11	76	NS	16.0
S. epidermidis	7	19	15	41	NS	
Peptostreptococcus	0	1	0	1	NS	0.2
Peptococcus	0	2	0	2	NS	0.4
Pseudomonas	2	25	0	27	<0.001	5.7
Candida	0	2	0	2	NS	0.4
Torulopsis	0	2	0	2	NS	0.4
CDC group IIIA	1	8	4	13	NS	

[a]Total = 642.
[b]By chi square analysis, for difference between media.
[c]Based on total positive minus 166 presumed contaminants equals 475.

From Hall, M., Warren, E., and Washington, J. A., II, *Appl. Microbiol.,* 27, 187, 1974. With permission.

TABLE 5

Comparison of TSB, Thiol Broth, and Aerobic Vacutainer® Culture Tubes With Supplemented Peptone Broth

					Supplemented peptone broth in Vacutainer tubes				
	TSB		Thiol		Aerobic		Anaerobic		
Organism	No. positive	Days (mean)	No. positive	Days (mean)	No. positive	Days (mean)	No. positive	Days (mean)	p[a]
Corynebacterium[b]	51	8.2	13	12.5	13	8.4	11	7.4	<0.01
Escherichia coli	87	2.0	81	1.8	34	2.1	36	1.8	<0.01
Enterobacter	12	1.0	12	1.9	2	1.0	3	1.3	<0.01
Haemophilus	4	9.8	2	12.0	0		0		Footnote c
Streptococci									
Viridans	17	1.8	14	2.4	6	1.7	7	1.6	<0.01
Group A	15	1.7	13	2.7	6	1.3	10	1.5	<0.01
Group D	6	1.5	6	1.3	2	2.0	3	1.7	Footnote c
Alcaligenes	8	2.3	1	2.0	2	5.0	1	3.0	Footnote c
Bacteroidaceae	24	2.3	26	2.5	9	3.4	8	3.4	<0.01
Staphylococci									
S. aureus	52	2.9	38	4.1	23	3.1	20	3.6	<0.01
S. epidermidis	41	4.5	30	5.4	17	3.2	16	4.2	<0.01
Pseudomonas	28	4.0	3	6.7	12	3.9	7	5.4	<0.01
Candida	6	3.2	2	4.0	2	4.5	1	2.0	Footnote c

[a]For hypothesis that proportions of positives are the same in all four media.
[b]Includes *Propionibacterium.*
[c]Although $p < 0.05$ in these instances, the sample sizes were too small for determination of significance.

From Hall, M., Warren, E., and Washington, J. A., II, *Appl. Microbiol.*, 27, 187, 1974. With permission.

soy broths, Rosner[47] encountered significantly fewer positive cultures in the culture tube than in each of the bottles. While it is tempting to conclude that supplemented peptone broth is a less adequate blood culture medium than those with which it was compared, an obstacle to this interpretation, which has already been discussed, is the volumes of blood cultured in each system. In the study by Hall et al.,[46] each culture tube received 2 ml of blood, and each bottle received 10 ml of blood. In Rosner's study,[47] each bottle received 5 ml of blood compared to the 2 ml inoculated into the tube. The effects of the volume of blood cultured on the rate of recovery of bacteria, as established by Hall et al.[38] using TSB and by Tenney et al.[39] using supplemented peptone broth, are significant. The comparisons between supplemented peptone broth and the other media were, in fact, comparisons between blood culture systems rather than between media. In view of the findings by Tenney et al.[39] that there was an average increase in the detection of clinically significant bacteria of 4.5%/ml, it is likely that supplemented peptone broth would have performed significantly better had larger volumes of blood, i.e., 5 or 10 ml, been cultured in larger volumes of this medium. This thesis is supported to some extent by the results obtained in a study by Painter and Isenberg[48] who compared the performance of the 50-ml plugged and vented Vacutainer culture bottles containing supplemented peptone broth with that of 50 ml of Eugon broth containing SPS, and of 50 ml of TSB containing SPS, yeast extract, and cooked liver. The Eugon broth and TSB bottles were produced in their laboratory, were not under vacuum, and contained no CO_2. Their analysis was based on 73 positive sets of clinical blood cultures and 132 positive sets of post-mortem heart blood cultures. They found that the Vacutainer bottles yielded *Bacteroides fragilis* significantly more frequently ($p < 0.05$) than did their in-house bottles, and that the latter yielded nonfermenting Gram-negative bacilli and yeasts significantly more frequently ($p < 0.05$) than the former. It is doubtful that the clinical blood culture and heart blood culture data are comparable and, therefore, can be analyzed together for two reasons. First, barring the use of some fairly vigorous efforts, it is very uncommon to be able to obtain 20 ml of heart blood post-mortem, and it is likely that few of these culture sets contained more than 10 ml of heart blood in the aggregate. Second, there are no data regarding the order of magnitude of the bacteremia in heart blood and its variability related to the degree to which the prosector compresses the abdomen or exerts pressure on other parts to try to obtain more blood. In examining the clinical and post-mortem data separately, therefore, the numbers of positive cultures by organism group are generally too small for analysis, and the differences cited above do not seem significant. The fact remains, however, that all three of these media performed comparably.

In 1969, Morello and Ellner[49] described a new medium – Columbia broth – for blood cultures. Its performance was compared with that of TSB by splitting blood samples sent to the laboratory in a sterile evacuated blood collection tube containing SPS and inoculating equal volumes into each of the media. Of the 99 species isolated, there were no significant differences in isolation rates between the two media; however, earlier growth in the Columbia broth was detected in a significant ($p < 0.01$) number of instances. Since 1969, Columbia broth has become widely used. In 1973, it was compared with TSB in a parallel study of 6904 blood cultures.[50] Both bottles contained 100 ml of the media with SPS under vacuum with CO_2. Routine subcultures were made after 24 hr of incubation, and the bottles were examined daily for 14 days. The two media performed equally well (Table 6) with only three exceptions: *Bacillus* was isolated more frequently ($p < 0.01$) from Columbia broth, and *S. aureus* and *P. aeruginosa* were isolated more frequently ($p < 0.01$ and p 0.05, respectively) from TSB. Contrary to the Morello and Ellner[49] study, there were no statistically significant differences between the two media in the time intervals required for the detection of growth (Table 7). The reasons for this variation in results are not clear. The sample size in the Morello and Ellner[49] study was small, the concentrations of SPS in the media during the period of study varied, the blood was transported to the laboratory in an evacuated blood collection tube containing SPS, and the media were not bottled under vacuum with CO_2. A problem with the isolation of *S. aureus* was suggested in the study, because, of 11 such strains isolated, 6 grew in both media, 4 in TSB only, and 1 in the Columbia broth only.

Another commonly employed blood culture medium is brain heart infusion broth (BHI) which

TABLE 6

Numbers of Isolates in Positive Cultures, by Medium

Organism	TSB and Columbia	TSB only	Columbia only	Total positive	p[a]
Bacillus	0	7	24	31	<0.01
Clostridium	6	0	2	8	NS[b]
Corynebacterium	23	55	40	118	NS
Lactobacillus	2	0	1	3	NS
Escherichia	79	27	21	127	NS
Salmonella	0	0	1	1	NS
Citrobacter	2	0	0	2	NS
Klebsiella	37	5	5	47	NS
Enterobacter	2	2	1	5	NS
Serratia	10	1	0	11	NS
Proteus	11	8	7	26	NS
Haemophilus	10	1	0	11	NS
Listeria	2	0	2	4	NS
Streptococcus					
S. pneumoniae	3	4	2	9	NS
Viridans	36	5	4	45	NS
Group A	0	0	1	1	NS
Group B	0	2	2	4	NS
Group D	19	5	1	25	NS
Eubacterium	0	1	0	1	NS
Acinetobacter	0	2	1	3	NS
Alcaligenes	0	1	3	4	NS
Flavobacterium	1	0	0	1	NS
Bacteroidaceae	18	11	15	44	NS
Micrococcus	0	0	1	1	NS
Staphylococcus					
S. aureus	60	25	5	90	<0.01
S. epidermidis	25	33	19	77	NS
Peptostreptococcus	3	1	0	4	NS
Peptococcus	0	0	1	1	NS
Veillonella	1	0	0	1	NS
Pseudomonas	35	15	6	56	0.05
Aeromonas	3	1	1	5	NS
Candida	6	1	1	8	NS
Torulopsis	0	3	0	3	NS

[a]For difference between media.
[b]NS, not significant.

From Hall, M., Warren, E., and Washington, J. A., II, *Appl. Microbiol.*, 27, 699, 1974. With permission.

was recently compared in parallel with TSB. Both bottles (Difco Laboratories, Detroit, Michigan) contained 100 ml of media under vacuum with CO_2 and were transiently vented upon receipt in the laboratory. Both were routinely subcultured on the day of their receipt in the laboratory and 48 hr later, and both were examined daily for 7 days and then again after 14 days of incubation. There were no statistically significant differences between the two media in the isolation rates of bacteria and yeasts (Table 8) or in the time interval required to detect growth.

In 1947, Castaneda described a bottle containing biphasic medium for the isolation of brucellae from blood.[51] The bottle contained an even, transparent layer of pancreatic digest of casein in agar on one sidewall and 10 ml of the same basal medium in liquid form. This medium differs from TSB in that it lacks soybean peptone. In 1951, this technique was employed by Scott[52]

TABLE 7

Time Intervals to Detection of Positivity

Organism	TSB		Columbia	
	No.	Mean ± SD[a] (days)	No.	Mean ± SD (days)
Bacillus	7	3.9 ± 4.6	24	4.5 ± 2.7
Clostridium	6	1.0 ± 0	8	1.0 ± 0
Corynebacterium	78	8.2 ± 3.1	63	9.3 ± 4.0
Lactobacillus	2	6.5 ± 0.7	3	5.0 ± 2.6
Escherichia	106	1.6 ± 1.6	100	1.6 ± 1.5
Salmonella	0		1	1.0
Citrobacter	2	1.0 ± 0	2	1.0
Klebsiella	42	2.3 ± 2.7	42	1.9 ± 2.1
Enterobacter	4	1.0 ± 0	3	1.0 ± 0
Serratia	11	1.2 ± 0.4	10	1.5 ± 1.0
Proteus	19	1.9 ± 1.2	18	2.1 ± 1.6
Haemophilus	11	2.5 ± 1.5	10	3.7 ± 4.4
Listeria	2	2.0 ± 0	4	2.0 ± 0
Streptococcus				
S. pneumoniae	7	1.3 ± 0.5	5	1.6 ± 0.5
Viridans	41	2.1 ± 1.8	40	1.9 ± 1.3
Group A	0		1	2.0
Group B	2	1.0 ± 0	2	1.5 ± 0.7
Group D	24	2.3 ± 1.6	20	2.4 ± 1.6
Eubacterium	1	7.0	0	
Acinetobacter	2	2.0 ± 0	1	2.0
Alcaligenes	1	7.0	3	9.6 ± 5.7
Flavobacterium	1	5.0	1	7.0
Bacteroidaceae	29	4.5 ± 3.7	33	3.8 ± 3.5
Micrococcus	0		1	5.0
Staphylococcus				
S. aureus	85	2.9 ± 2.9	65	1.9 ± 1.2
S. epidermidis	58	3.8 ± 2.1	44	3.9 ± 2.2
Peptostreptococcus	4	1.5 ± 0.6	3	1.3 ± 0.6
Peptococcus	0		1	7.0
Veillonella	1	3.0	1	3.0
Pseudomonas	50	3.5 ± 2.6	41	3.1 ± 1.7
Aeromonas	4	1.3 ± 0.5	4	1.5 ± 0.6
Candida	7	8.3 ± 2.6	7	7.3 ± 0.9
Torulopsis	3	4.3 ± 1.5	0	

[a]SD, standard deviation.

From Hall, M., Warren, E., and Washington, J. A., II, *Appl. Microbiol.*, 27, 699, 1974. With permission.

to prepare a Castaneda bottle containing soybean-casein digest agar and broth for routine blood culture purposes. More recently, Coetzee and Johnson[53] have incorporated 2-, 3-, 5-triphenyltetrazolium chloride in the agar slant to facilitate the detection of bacteria. Although the authors conclude that an advantage of this system is the saving of time and media by not having to perform routine subcultures, they provide no data comparing detection rates of bacteria in this system with those in a broth medium which is routinely subcultured. The theoretical advantages of the Castaneda principle for routine bacterial blood cultures are certainly attractive but still require controlled studies.

While it is difficult to segregate the effects of media from those of the atmosphere of the incubation of cultures on the isolation of anaerobic bacteria, it is appropriate at this time to examine those studies of blood culture media which pertain to anaerobes. The isolation of anaerobes from the blood is of considerable

TABLE 8

Numbers of Isolates in Positive Cultures, by Medium

Organism	TSB and BHI	TSB only	BHI only	Total positive
Bacillus	0	0	2	2
Corynebacterium	7	9	11	27
Escherichia	41	11	14	66
Citrobacter	2	0	0	2
Klebsiella	11	0	2	13
Enterobacter	1	2	1	4
Serratia	2	3	2	7
Proteus	8	1	3	12
Haemophilus	4	0	1	5
Streptococcus				
S. pneumoniae	4	2	1	7
Viridans	11	2	3	16
Group B	1	0	0	1
Group D		3	1	4
Acinetobacter	1	0	0	1
Flavobacterium	1	0	1	2
Neisseria	2	0	0	2
Bacteroidaceae	8	3	1	12
Staphylococcus				
S. aureus	16	4	3	23
S. epidermidis	4	7	16	27
Pseudomonas	15	5	3	23
Candida	5	0	2	7
Cryptococcus	0	0	1	1
Total	144	52	68	264

importance because of their known frequency of occurrence and clinical significance.[54] It has been commonly assumed that thioglycollate broth provides the optimal opportunity for the isolation of anaerobes from blood because its E_h is commonly in the range of −0.175 to −0.200 V.[55] The E_h of Thiol is also apparently in this low range.[135] Nonetheless, the studies comparing TSB with thioglycollate[45] and with thiol[44,46] have failed to demonstrate any advantages of these reduced media in the isolation of anaerobic bacteria from blood. In 1973, Washington and Martin reported a study comparing TSB with thioglycollate medium and supplemented, prereduced BHI for the recovery of anaerobic bacteria from blood[56] and found no significant differences among these three media in the number of anaerobes detected (Table 9) or in the time intervals required to detect them (Table 10).

A survey in England by Shanson[57] to determine methods used for anaerobic blood cultures showed that approximately one third of the laboratories used cooked meat medium, one quarter used thioglycollate, another one third used no specific anaerobic medium, and the remainder used a variety of methods. Simulated blood cultures were performed with several strains of clostridia, anaerobic cocci, and anaerobic Gram-negative bacilli to determine the ability of several different types of media, with and without SPS, to yield growth of small inocula. In general, thioglycollate without SPS provided the earliest and most reliable isolation of nonsporulating anaerobes, although a glucose cooked-meat digest broth with SPS also gave good results. One particular formulation of thioglycollate (Southern Group Laboratories Brewer's) performed somewhat better than the others, possibly because its pH was more alkaline initially (7.5) and because its glucose concentration of 0.25% was half that of the others tested (Oxoid Brewer's and USP). It was speculated that any fermentation and acidity resulting from bacterial growth in this medium might not be as detrimental to bacterial survival as that in the other formulations of thioglycollate medium. Further exploration of this issue with clinical specimens is certainly indicated.

Forgan-Smith and Darrell[58] compared five anaerobic media containing SPS in a model blood culture system. The media studied were USP thioglycollate medium, thioglycollate medium (Brewer's), reinforced clostridial medium, dehydrated cooked-meat medium, and freshly made cooked-meat medium. Each was inoculated with either fewer than 100 or fewer than 10 colony-forming units of anaerobic Gram-negative bacilli. The USP thioglycollate medium was found to yield significantly ($p < 0.05$) better isolation rates than the other media studied.

In another model blood culture study, Szawatkowski[59] compared Thiol medium with USP thioglycollate medium, cooked-meat medium, and glucose broth by inoculating small numbers of *Bacteroides* in triplicate into each medium. He found that Thiol medium yielded more positive cultures than did the other media, and that all strains were more rapidly isolated in Thiol medium than in the other media.

It seems safe to conclude from these simulated blood culture studies that Thiol and thioglycollate media were superior to the other media tested in the isolation of small inocula of anaerobic Gram-negative bacilli. This superiority remains to be demonstrated, however, in clinical blood culture studies. One disadvantage of thioglycollate medium mentioned in all three reports was the

TABLE 9

Numbers of Anaerobic Isolates in Positive Cultures

	No. positive					
	By medium[a]			By combination of media[a]		
Organism	TSB	Thio	BHI	TSB + Thio	Thio + BHI	TSB + Thio + BHI
Bacteroides fragilis	18	19	19	1	1	15
B. melaninogenicus	2	1	1			1
Fusobacterium fusiforme	1					
Eubacterium cylindroides	1					
Peptococcus	1	1	1			1
Peptostreptococcus	1	1		1		

[a]TSB, tryptic soy broth; thio, thioglycollage; BHI, brain heart infusion.

From Washington, J. A., II and Martin, W. J., *Appl. Microbiol.*, 25, 70, 1973. With permission.

TABLE 10

Time Interval to Positivity for *Bacteroides fragilis*

Medium	No.	Days, mean
TSB	18	3.1
Thio	19	2.7
BHI	19	2.5

From Washington, J. A., II and Martin, W. J., *Appl. Microbiol.*, 25, 70, 1973. With permission.

lack of continued viability of anaerobic Gram-negative bacilli over a period of several days; as a storage medium, cooked meat provided the longest viability of cultures.

Lack of viability of certain bacteria in TSB was noted by Waterworth.[61] She placed small inocula of several bacterial species into this medium and compared their viability with that in BHI. In general, the pH in TSB fell below 6.0 following bacterial growth, while that in BHI remained above this level. In these studies, four strains of *Vibrio cholerae* did not survive overnight in TSB, and several strains of pneumococci either failed to grow in this medium or were recovered in small numbers. Since several strains of pneumococci also failed to grow in BHI, Waterworth concluded that the suitability of these media for blood cultures was open to question. The significance of these findings, once blood and its buffering capacity have been added to the media during clinical use, is uncertain and bears further scrutiny. It seems quite possible that the seriousness of this problem, as will be discussed later in this chapter, is related to the timing of routine subcultures.

It is difficult to ascribe an increase in anaerobic bacteremias to the type of medium used for blood cultures. As shown in Figure 3 in the number of patients with Gram-negative bacillemias at the Mayo Clinic and its affiliated hospitals between 1950 and 1975, there was a marked increase in those caused by *Bacteroides* between 1967 and 1968. This increase was clearly due to the replacement of the dextrose brain broth media used at the Mayo Clinic for blood cultures until 1968 by TSB and thiol medium under vacuum with

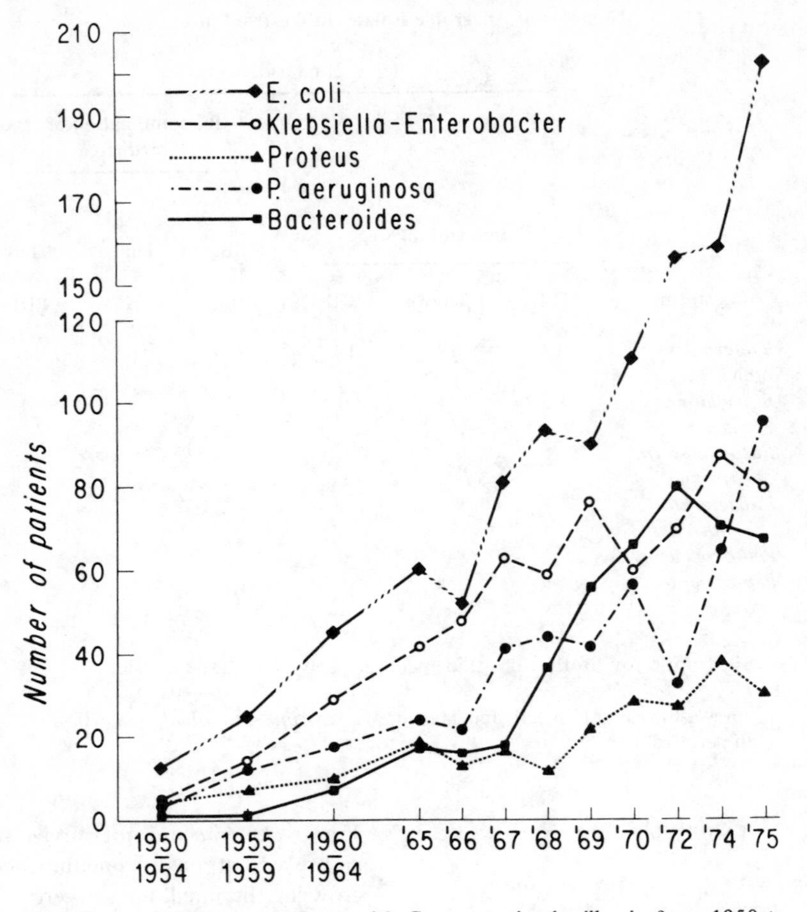

FIGURE 3. Number of patients with Gram-negative bacillemia from 1950 to 1975, Mayo Clinic and affiliated hospitals.

CO_2.[60] This basic system continued in use until 1973, at which point SPS was included in these media. Yet it is apparent in Figure 3 that the number of patients with *Bacteroides* bacteremias continued to increase markedly until 1972. There were, obviously, other factors during this interval which accounted for the increase in the number of these patients.

Based on Mayo Clinic data and an analysis of published reports of clinical blood culture studies, it would appear that either TSB or BHI is most suitable as routine blood culture media. This recommendation must be tempered, however, by reemphasizing the fact that the majority of the Clinic's studies have compared bottles of media produced by the same manufacturer, so that variables, other than those related to the basic composition of the media per se, have probably been minimal. This point was recently reinforced by studies of Mangels et al.,[136] who found another manufacturer's BHI broth bottle (Pfizer Diagnostics Division, New York) to yield significantly fewer anaerobic isolates than thiol or supplemented peptone broth. We have been unable to document that thioglycollate and thiol media provide any better recovery of anaerobic bacteria from blood than TSB; however, it is clear that these media do severely impair the recovery of *P. aeruginosa,* despite routine subcultures, and should not, therefore, be relied upon completely as blood culture media.

Volume

Kracke and Teasley[62] noted in 1930 that, although there was no uniformity in reported blood culture methods, it was evident that the quantity of blood used was small and the volume of medium inoculated was large. They conducted a series of experiments which showed that the optimum dilution of blood in media was approximately 1:20. They ascribed the bactericidal activity of blood, diluted less than 1:20 in medium, to complement and studied ways by which complement could be bound or inactivated. They found peptone to have relatively slight complement-fixing activity, but also found that fine meat particles in suspension exhibited a high level of complement fixation. Ultimately, they described a medium containing meat extract and a suspension of brain tissue for blood cultures. These investigators' studies are, of course, of interest in terms of their application of the significance of complement and its fixation to blood culture procedures.

Although our knowledge about the factors responsible for the bactericidal activity of blood has become infinitely more complex over the years, the simple fact remains that blood must be diluted in culture medium to reduce its bactericidal activity. It is generally recommended that this dilution be in the range of 1:10 to 1:20, despite the fact, recognized by Kracke and Teasley,[62] that different media and their components may vary in their ability to neutralize the bactericidal activity of blood.

Roome and Tozer[63] placed small inocula of a variety of bacteria into blood-nutrient broth mixtures in ratios ranging from 1:1 to 1:60. They found that the bactericidal activity of normal blood was nearly always prevented by its 1:30 dilution in broth. Lowrance and Traub[64] showed that 50% fresh serum killed more than 10^4 organisms per milliliter within 5 min and that at least 20 to 40 min were required for 20 and 10% fresh serum, respectively, to kill a comparable inoculum. They found that none of the liquid media tested (nutrient, Mueller-Hinton, thioglycollate, and TS) depressed the bactericidal activity of serum.

Based on their studies, Roome and Tozer[63] stated that the minimum volume of broth necessary for culture of 5 ml of blood is 150 ml and recommended that a dilution of 1:60, i.e., 5 ml of blood into 300 ml of medium, be used. Obviously, dealing with bottles or flasks which are sufficiently large to contain these volumes of media poses a considerable problem. It was, therefore, fortunate that these two studies were published at a time when the addition of SPS to blood culture media was becoming popular among commercial manufacturers of these products.

Additives

Although virtually any component of broth beyond its basal medium might be considered to be an additive, it is more realistic to confine this discussion to the following categories: anticoagulants and antibiotic neutralizers, sucrose and other agents designed to produce hypertonicity of the medium, and cysteine.

By 1938, when Von Haebler and Miles[65] reported the results of their investigations of the action of sodium polyanetholsulfonate (SPS or Liquoid®), this polyanion had been used as an additive to blood culture media in Europe for 5 years. They compared the growth of *Streptococcus pneumoniae, Brucella melitensis, Haemophilus influenzae,* and *Staphylococcus aureus* in model blood culture systems supplemented with three anticoagulants – sodium citrate, trypsin, and SPS – and found that sodium citrate partially inhibited *B. melitensis* and completely inhibited as many as 1.5×10^5 *H. influenzae;* that trypsin partially inhibited *H. influenzae* and completely inhibited 1.9×10^3 *S. pneumoniae;* and that SPS in concentrations of 0.1 and 0.5%, if anything, stimulated the growth of *S. aureus, S. pneumoniae,* and *B. melitensis* but markedly inhibited *H. influenzae* at these concentrations. They also found that meningococci were inhibited by all but 0.05 and 0.025% SPS. At 0.05%, a large number of strains tested, including *H. influenzae, S. typhi,* and *Bacteroides,* grew without difficulty. Most of the previously reported work with SPS, reviewed by Von Haebler and Miles,[65] had used higher concentrations (0.17%) of this substance and had found it to be inhibitory to certain organisms. Von Haebler and Miles,[65] therefore, recommended the use of 0.03 to 0.05% SPS in routine blood cultures.

Penfold et al.[66] found that, although a variety of bacteria tested did grow in 0.17% SPS in saline, small inocula of two strains of viridans streptococci did not and concluded that SPS in broth or in solution gave disappointing results. Garrod[67] subsequently tested 14 strains of viridans streptococci in 0.03% SPS and found that not only were none inhibited, but also that all could be recovered in medium containing SPS when the dilution of blood in medium was 1:2.

In 1939, Hoare[68] examined the effects of SPS on anaerobic streptococci in a model blood culture system and found that media containing 0.03% SPS yielded scanty or no growth of these organisms within 96 hr of incubation. He concluded that, since anaerobic cocci were frequently present in the blood in puerperal fever, the routine addition of SPS to blood culture media was not recommended. The selective inhibition in vitro of *Peptostreptococcus anaerobius* by SPS has been confirmed by Graves et al.,[69] who proposed that a disc containing the substance could be used as a presumptive test for the identification of this species. These findings were confirmed by Kocka et al.[70] in a study comparing the effects on anaerobic cocci of SPS and another polyanion, sodium amylosulfate (SAS).

Although these in vitro effects of SPS on *P. anaerobius* are impressive, the fact remains that anaerobic streptococci generally are recovered more frequently from blood culture media with SPS than from media without it,[71] suggesting either that *P. anaerobius* is a rare cause of anaerobic bacteremia, or that model blood culture systems inaccurately reflect what goes on in clinical blood culture systems. In fact, both suppositions may be correct. In a recent study of 5800 blood cultures at the Mayo Clinic in which TSB containing SPS was compared to TSB containing SAS, which is not inhibitory in vitro to *P. anaerobius,*[70] there were no isolates of this species out of 392 bacteria recovered.[72] Wilkins and West[73] found that the sensitivity of *P. anaerobius* to SPS was medium dependent, confirming an observation made previously by Shanson.[57] They demonstrated that gelatin, proteose peptone, and casein protected *P. anaerobius* from the toxicity of SPS. These substances are present in varying amounts in supplemented peptone broth, BHI, and TSB.

That the addition of SPS to culture media increases the rates of isolation of various bacteria from blood is beyond question. Parallel studies by Rosner,[71] Finegold et al.,[74] and Eng[75] have shown the strikingly beneficial results provided by media containing SPS. Rosner's studies[71] and those of Eng[75] demonstrated that SPS improved the recovery of Gram-positive cocci and Gram-negative bacilli. Eng[75] also showed that SPS increased the rapidity with which growth could be detected in broth. A notable exception to these findings was *Neisseria meningitidis,* growth of which Eng[75] found to have been markedly inhibited in broth containing 0.05% SPS. Eng and Iveland[76] subsequently confirmed these findings in vitro and also found similar degrees of inhibition by SPS against *N. gonorrhoeae.* Just why their results differed from those reported by Von Haebler and Miles[65] is unclear, but may be related to the small number of strains tested in the 1938 studies or possibly to strain variations in sensitivity to SPS. At any rate, Eng and Iveland[76] concluded that media with and without SPS should be used routinely. This poses a serious dilemma for the laboratory, because the isolation rates of bacteria other than pathogenic Neisseriae are significantly higher in media containing SPS. The laboratory, therefore, must either provide media with and without SPS for routine use or must try to impose a selective system of cultures whereby blood is inoculated into a medium without SPS only when meningococcemia or gonococcemia is suspected on clinical grounds. Since the predictability of these particular clinical entities is quite high, the latter procedure seems the better. The blood culture order form used by the Mayo Clinic's clinical staff lists *Neisseria* separately, and blood for culture of these organisms is inoculated into a bottle with medium which does not contain SPS. A Castaneda bottle containing TS agar and broth represents a convenient and effective system for this purpose.

The net effect of the addition of SPS to culture media is to reduce the dilution of blood required to overcome its bactericidal effect. Lowrance and Traub[64] showed that 0.0125 to 0.025% SPS was sufficient to neutralize the bactericidal activity of 50% fresh human serum, and that 0.006% SPS neutralized the bactericidal activity of both 10 and 20% fresh human serum; however, they also showed that this neutralizing property occurred only when SPS was added to serum prior to the addition of the organisms. Belding and Klebanoff[77] examined the effects of SPS on serum bactericidal activity, on the phagocytic and microbicidal activity of isolated leukocytes, and on certain leukocytic metabolic parameters. By inoculating 20 or fewer organisms representing 15 different species into whole blood with and without 0.05% SPS, they demonstrated that 14 of the species were recovered significantly ($p < 0.01$) more frequently from the blood containing SPS. As little as 0.001% SPS decreased, but did not abolish, the serum bactericidal effect. SPS at a final concentration of 0.05% also abolished the microbicidal activity of leukocytes due to inhibition of serum-mediated phagocytosis of the test organisms. Finally, these investigators demonstrated that SPS stimulated glucose C-1 oxidation by resting human leukocytes.

As was previously noted, Kracke and Teasley[62] described a blood culture medium which contained beef extract and brain tissue and which exhibited a high level of complement fixation. Kracke's blood culture medium, which is obtainable commercially, was tested in parallel with TSB containing 0.025% SPS in a limited study of 1780 blood cultures at the Mayo Clinic. While half of the 140 positive cultures were detected in both media, 48 were detected in TSB only and 22 in Kracke's medium only (Table 11). An obvious disadvantage of Kracke's medium is its inherent turbidity, rendering macroscopic examination quite difficult and necessitating the more frequent use of stained smears and subcultures. Be that as it may, the theoretical advantages of Kracke's medium have probably been offset by the addition of SPS to a clear broth medium.

Studies by Kocka et al.[78] on the action of sulfated polyanions showed that these substances inhibited the activity of lysozyme, inactivated complement, and decreased the antibacterial activity of streptomycin, kanamycin, gentamicin, and neomycin. Medium-dependent antagonism of gentamicin by SPS had been previously noted by Traub and Lowrance[79] and has been utilized as a means of selectively inhibiting aminoglycosides and polymyxins in antibiotic assays.[80]

In 1972, Kocka et al.[81] described a new anticoagulant – (SAS) – the properties of which closely resembled those of SPS[78] with the exception that it did not interfere with the growth of *P. anaerobius.*[70] In limited clinical blood culture trials, Kocka et al.[82] showed that the presence of 0.05% SAS in the medium increased the number of positive cultures compared to those in media without SAS, and that the

TABLE 11

Number of Isolates by Medium

	TSB(v) only[a]	Kracke's-TSB(v) only[a]	Both TSB(v) and Kracke's-TSB(v)[a]
Bacillus	1	1	
Clostridium septicum			1
Corynebacterium	1		
Propionibacterium acnes	2	3	
Escherichia coli	8	7	21
Citrobacter freundii			2
Klebsiella pneumoniae	2	1	3
Enterobacter aerogenes		1	2
E. cloacae	2		1
Serratia marcescens	2	1	1
Proteus mirabilis	1	2	5
Haemophilus influenzae	1		2
Streptococcus pneumoniae	8		3
Streptococcus, D		1	3
Streptococcus, viridans	2	1	8
Eubacterium lentum			2
Acinetobacter calcoaceticus	1		
Bacteroides fragilis	6	2	3
Staphylococcus aureus	1		3
S. epidermidis	3	1	5
Peptococcus			2
Other anaerobic cocci	2		
Pseudomonas aeruginosa	1	1	3
P. fluorescens	1		
Candida albicans	3		
Total	48	22	70

[a]v = transiently vented.

number of organisms isolated from thioglycollate medium with SAS was comparable to that in the same medium with SPS. Somewhat different results were reported by Hall et al.[72] who compared TSB containing 0.025% SPS with TSB containing initially 0.05% SAS and later 0.025% SAS. Since there were no differences between the 0.025 and 0.05% SAS, the results were combined (Table 12). Although there were no statistically significant differences among the organism groups isolated between SPS and SAS, there were more isolates, and particularly more Gram-negative bacilli, recovered from TSB containing SPS than from TSB containing SAS. Moreover, there was no dramatic increase in the number of anaerobic cocci encountered, nor was a single isolate of *P. anaerobius* recovered. Finally, SAS imparted a faint turbidity to the TSB which slightly hindered macroscopic examination of the bottle. At this point, therefore, SAS seems not to have imparted any particular advantages relative to SPS to clinical

TABLE 12

Isolation Rates in TSB Containing SPS and SAS

Organism	Both	SPS only	SAS only
Corynebacterium	8	20	16
Clostridium	0	1	0
Escherichia	53	21	15
Citrobacter	0	1	2
Klebsiella	13	7	1
Enterobacter	8	1	1
Serratia	7	2	3
Proteus	12	0	2
Providencia	1	0	0
Haemophilus	2	5	0
Streptococcus			
S. pneumoniae	3	0	1
Viridans group	12	5	3
Other	13	2	1
Eubacterium	1	0	0
Acinetobacter	1	0	1
Alcaligenes	0	2	0
Bacteroidaceae	16	8	7
Staphylococcus aureus	26	12	9
S. epidermidis	11	15	7
Peptostreptococcus	2	0	3
Peptococcus	0	0	2
Veillonella	0	1	3
Pseudomonas	15	8	3

From Hall, M. M., Warren, E., Ilstrup, D. M., and Washington, J. A., II, *J. Clin. Microbiol.*, 3, 212, 1976. With permission.

blood cultures, but it has some very definite disadvantages which should limit its utility considerably.

Despite the fact that most investigators have recommended that SPS be incorporated in media at a concentration of 0.025%, Rosner[83] has raised the question as to whether or not 0.05% should be used. In a parallel study of flasks containing TSB with 10% sucrose and 0, 0.025, 0.05, and 0.075% SPS, fewer organisms were recovered from the media containing 0.025 and 0.075% SPS than from that containing 0.05%. Since the selection of the lower concentration was based to a great extent on the premise that it was less likely to be inhibitory to anaerobic streptococci than would be 0.05%, and since this problem is more hypothetical than real in clinical blood cultures, Rosner's recommendation may have merit, especially when one is using hypertonic media. The degree to which his findings apply to nonhypertonic media is, of course, not known at this time. Naturally, his study once again confirmed the value of SPS in blood culture media.

While the value of SPS seems rather clear-cut, and few would argue about its use as an additive in blood culture media, the literature on the value of hypertonic media is far less straightforward. Conflicting results have been reported for reasons that remain to be satisfactorily explained.

In 1972, Rosner[71] performed a parallel clinical blood culture study of three flasks with Brucella broth containing no additives, 0.05% SPS, and both 0.05% SPS and 30% sucrose. His results (Tables 13 and 14) demonstrate the beneficial results of SPS, especially in those bacteremias of a low order of magnitude, and the added beneficial effect of 30% sucrose. The overall numbers of bacteria recovered by organism group and by medium are listed in Table 14 and again demonstrate marked differences in rates of recovery between Brucella broth without SPS and the same medium with SPS. There were additional isolates in the Brucella broth with both SPS and sucrose. In a later study comparing a system (E-Vac,® Pfizer Diagnostics, New York) with and without 10% sucrose, Rosner[84] recovered 356 and 319 organisms, respectively, but found that many organisms tended to die off more rapidly in the hypertonic than in the isotonic medium. The basal medium in the system and whether SPS was included were not specified in this report.

Henrichsen and Bruun[85] performed a study in

TABLE 13

Number of Isolates in Positive Cultures, by Medium and According to Order of Magnitude of Bacteremia

Colony-forming units/ml	Brucella broth		
	No additives	0.05% SPS	0.05% SPS, 30% sucrose
<20	1	16	22
21–100	22	31	39
>100	51	56	60

Adapted from Rosner, R., *Am. J. Clin. Pathol.*, 57, 220, 1972.

TABLE 14

Number of Isolates in Positive Cultures, by Medium and Organism Group

Organism	Brucella broth		
	No additives	0.05% SPS	0.5% SPS, 30% sucrose
Streptococcus			
Alpha hemolytic	37	41	41
Beta hemolytic	10	12	12
Anaerobic	0	6	9
S. pneumoniae	24	34	34
Staphylococcus aureus	3	3	3
Neisseria meningitidis	0	2	5
Haemophilus	0	1	4
Pasteurella	0	1	2
Bacteroides	0	6	11

Adapted from Rosner, R., *Am. J. Clin. Pathol.*, 57, 220, 1972.

which 8 ml of blood was transported to the laboratory in an evacuated tube containing 1% SPS and was then distributed equally into each of 12 tubes. Four tubes contained a nutrient broth with serum and hemolyzed horse blood, four contained semisolid nutrient agar, and four contained semisolid thioglycollate agar (volumes unspecified). Half of the tubes with each medium contained 10% sucrose. The results are summarized in Table 15, and it is clear that the addition of sucrose to these media was helpful. In a more detailed analysis of isolates of *E. coli,* Henrichsen and Bruun[85] found that this species grew only in tubes containing sucrose in 43 instances and only in tubes without sucrose in 16 instances, a difference which was statistically significant ($p < 0.05$).

Sullivan et al.[86] compared the isolation rates of bacteria from TSB (under vacuum with CO_2 but without SPS), an anaerobic broth with 0.05% SPS which was osmotically stabilized with 16% sucrose and magnesium sulfate, and the same anaerobic broth without SPS and osmotic stabilizers. A sufficient volume of blood was also collected for a membrane filter culture and for five pour plates. The osmotically stabilized anaerobic medium with SPS yielded the largest number of isolates (Table 16). Of the 20 isolates recovered only in this medium, there were eight anaerobes and two protoplasts of *Propionibacterium acnes.* As impressive as these data are, they are somewhat difficult to assess because only osmotically stabilized medium contained SPS, and the relative importance of osmotic stability and SPS cannot be determined. The osmotically stabilized medium used in this study is highly complex and has not yet become available commercially.

TABLE 15

Number of Isolates by Organism Group and by Positivity in Media With and Without Sucrose

Organism	Growth more frequent in tubes with than without sucrose	Growth less frequent in tubes with than without sucrose
Escherichia coli	63[a]	40
Klebsiella	10	7
Proteus mirabilis and *P. vulgaris*	10[a]	3
Other Enterobacteriaceae	7	7
All Enterobacteriaceae	90[b]	57
Pseudomonas aeruginosa	8	3
Staphylococcus epidermidis	30	27
Staphylococcus aureus	31[a]	18
Streptococci	10	8
Bacteroides	3	5

[a] $p < 0.05$.
[b] $p < 0.01$.

Adapted from Henrichsen, J. and Bruun, B., *Acta Pathol. Microbiol. Scand. Sect. B,* 81, 707, 1973.

TABLE 16

Number of Positive Cultures by Medium

Medium	Total positive	Earlier growth	Positive only in this medium
TSB	24	1	8
Anaerobic	27	2	9
Anaerobic, osmotically stabilized, and SPS	43	2	20

Adapted from Sullivan, N. M., Sutter, V. L., Carter, W. T., Attebery, H. R., and Finegold, S. M., *Appl. Microbiol.*, 23, 1101, 1972.

In 1975, Washington et al. reported a study of 5883 clinical blood cultures in which TSB, with and without 15% sucrose, under vacuum with CO_2 and containing 0.025% SPS was compared in parallel.[87] All bottles were routinely subcultured after 1 and 5 days of incubation. *Bacillus* was the only species isolated more frequently ($p < 0.01$) from the hypertonic TSB; however, *Haemophilus,* Bacteroidaceae and *S. aureus* were isolated significantly less frequently ($p < 0.05$) from the hypertonic TSB (Table 17). The mean time intervals required to detect growth were significantly shorter in TSB without sucrose for viridans streptococci ($p < 0.02$), Bacteroidaceae ($p < 0.05$), *S. aureus* ($p < 0.001$), and *S. epidermidis* ($p < 0.05$).

Chong et al.[88] found that 15% sucrose in BHI with SPS adversely affected the growth of *Salmonella typhi.* In concentrations of 10 to 20%, sucrose in BHI also diminished the growth in 24 hr of small inocula of *E. coli, P. aeruginosa, S. aureus,* viridans streptococci and *S. pneumoniae.*

The data in the latter two studies are difficult to reconcile with those cited earlier, and it remains unclear what the true value of hypertonic media is. The methods used in each of these studies varied, as did the media. One might speculate that the effects of added sucrose are medium dependent and, perhaps, even system dependent. Some support for this hypothesis can be found in a series of studies of hypertonic media reported by Ellner et al.[89] They compared isotonic and hypertonic Columbia broth, modified by the addition of cysteine, to a final concentration of 0.05% in a clinical study of 6000 specimens and found no clinically significant differences between the two media in vented bottles except for the more frequent recovery of *Pseudomonas* from the isotonic medium. In a second experiment, the two media in unvented bottles showed only that *S. pneumoniae* was recovered more frequently from the isotonic medium and *Klebsiella* more fre-

TABLE 17

Numbers of Isolates in Positive Cultures by Medium[a]

Organism	Both media	TSB with SPS only	TSB with SPS and sucrose only	p
Bacillus	1	7	23	<0.01
Clostridium	0	2	0	NS
Corynebacterium	17	32	37	NS
Escherichia	37	7	13	NS
Klebsiella	13	4	4	NS
Enterobacter	2	1	0	NS
Serratia	7	0	2	NS
Proteus	13	3	3	NS
Cardiobacterium	4	3	1	NS
Haemophilus	3	7	0	<0.05
Streptococcus				
S. pneumoniae	9	1	0	NS
Viridans group	22	5	5	NS
Group A	5	4	0	NS
Group D	6	1	3	NS
Other groups	0	0	1	NS
Alcaligenes	0	3	9	NS
Neisseria	0	1	0	NS
Bacteroidaceae	13	7	0	<0.05
Staphylococcus				
S. aureus	36	13	3	<0.05
S. epidermidis	13	22	11	NS
Peptostreptococcus	0	0	1	NS
Veillonella	0	1	0	NS
Pseudomonas	14	5	2	NS

[a]NS, not significant.

From Washington, J. A., II, Hall, M. M., and Warren, E., *J. Clin. Microbiol.*, 1, 79, 1975. With permission.

quently from the hypertonic medium; however, the numbers of positive cultures in any bottles with these two organisms were very small. In a third experiment in which two vented bottles containing isotonic media were compared, the authors found little difference in the rates of recovery of aerobes and anaerobes compared to the rates of recovery in the first experiment which were indicative of the value of culturing the additional volume of blood rather than of the hypertonic medium itself. Although slightly more facultatively anaerobic and anaerobic bacteria were isolated from the combination of bottles with isotonic and hypertonic media than from the duplicate bottles containing isotonic media, the difference was not statistically significant. Furthermore and most importantly, this comparison was made between two consecutive studies consisting of different numbers of samples and was not, therefore, statistically valid. Nonetheless, the authors concluded that the unvented hypertonic medium was better than the unvented isotonic medium for the recovery of facultatively anaerobic and anaerobic organisms and recommended its routine use, a conclusion and a recommendation which were not well supported by their data.

The hypothetical value of adding sucrose to blood culture media is that it provides osmotic stabilization for bacterial variants which are cell wall defective, either as the result of tissue, blood, or urine lysozyme activity or of antimicrobial therapy with antibiotics which interfere with cell wall biosynthesis. Sucrose may, in fact, do little more than inactivate any penicillin which may be present in the specimen, thereby permitting classical bacteria to grow only in the hypertonic medium.[90] It is apparent, however, that under certain as yet unspecified conditions, some bacteria may be recovered only from blood cultured in hypertonic media. The significance of such isolates has not been carefully scrutinized, but certainly should be according to strict criteria.[91]

The term wall-defective microbial variants is a general one which includes protoplasts, spheroplasts, transitional phase variants, L-phase variants, and unclassified wall-defective variants.[91] The requirement by such organisms for hypertonic media is not absolute, and they may survive osmolalities comparable to those of human serum and even distilled or deionized water.[91] Lorian and Waluschka,[92] for example, have reported the isolation of aberrant forms of Enterobacteriaceae from blood cultures in Thiol medium and TSB containing SPS from patients receiving antimicrobial agents. Phair et al.[93] performed multiple blood cultures of 28 patients with fever of unknown origin. Blood was inoculated into BHI and into BHI containing 0.2% $MgSO_4$ and 5% NaCl or 0.3 *M* (10%) sucrose. Since no mention was made of the presence of SPS in either medium, it is assumed that it was not added. All of these cultures were negative; however, cultures of another 14 patients taken during or following therapy for proven infective endocarditis, were negative in all but three cases, from whom cultures in the hypertonic media only were positive. The three organisms involved were *Candida parapsilosis, S. epidermidis,* and *S. aureus,* and one cannot help but wonder whether they might have been isolated had SPS been added to the conventional cultures. Nonetheless, the authors concluded that even in a selected population, the rate of isolation of wall-defective microbial variants was too low to warrant attempts at their routine cultivation.

Irwin et al.,[94] however, have advocated wider application of routine cultural surveillance for cell wall-deficient microbial variants after isolating such an organism, which ultimately proved to be a coryneform bacteria, from two blood cultures on an osmotically stabilized solid medium from a patient with a chronic febrile illness of unknown etiology. Conventional blood cultures in TSB with agar (Castaneda bottles?), but without SPS, remained negative. The osmotically stabilized medium, designated as L-phase growth medium (LGM), consisted of TSB with 1.5% agar, 3% ammonium chloride, and 2% horse serum. Two separate 1-ml samples of the patient's blood were inoculated onto this medium and yielded growth after 8 days. The patient, who was suspected of having right-sided infective endocarditis, responded after therapy with erythromycin (30 days), streptomycin (10 days), and, finally, tetracycline (30 days) and had no further positive cultures.

While there is little evidence to support these authors' contention that routine surveillance for wall-deficient microbial variants be applied more widely, one must concede the possibility that culture for such organisms under highly selective circumstances may be helpful. There are, however, real problems in interpreting reports, such as the one by Irwin et al.,[94] since we do not know the frequency with which these organisms might be recovered from a control or healthy population. In the absence of histopathological evidence of their presence in a lesion, both their origin and their relationship to the disease present remain questionable.

Louria et al.[95] studied specimens from 300 patients in several aerobic and anaerobic media, including 5 that were hypertonic. Only 0.5 ml of each specimen was inoculated into 10 ml of the several hypertonic media examined. In only 16 cases (5%) were organisms found in the hypertonic media which appeared to originate from the patient on the basis of repeated isolations, serologic studies, or both, and in only 9 of these were the organisms clearly considered to be clinically significant. In each of these instances, of which 7 represented isolates from blood, cultures in nonhypertonic media remained negative during the 7 to 10 days they were examined. Louria et al.[95] concluded, on the basis of this study and their earlier ones in which a total of 17 clinically significant isolates had been recovered from over 600 patients, that specimens for study in hypertonic media should be carefully selected from cases with clinically suspected meningitis or endocarditis and negative conventional cultures. They felt that since the amount of work involved in these studies was enormous and the yield low, the routine use of hypertonic media was not indicated.

Brogan[96] studied 1527 specimens of blood inoculated in parallel into conventional media and into one of several L-form media over a 2½ year period. L-form bacteria were isolated only from the media with an osmolality exceeding 1100 mOsm/kg in 6 cases; however, no clinical information about any of the cases is given in this paper.

The value of adding β-lactamase or, more specifically, penicillinase to blood culture media has not been carefully evaluated for many years. In 1945, Dowling and Hirsh[97] reported that when penicillinase was added to veal-infusion broth and kept at incubator, room, or refrigerator temperature, its potency did not diminish over the course of 4 weeks. Moreover, they found that the activity of penicillin in veal-infusion broth diminished only slightly over the course of 24 hr and that as little as 0.01 ml of penicillinase per unit of penicillin rapidly eliminated any detectable penicillin activity. Among 186 cultures collected from 26 patients receiving penicillin, growth was poor in 22 instances and absent in 14 instances in which the medium lacked penicillinase. Conversely, there was only one instance in which growth did not occur in the medium with penicillinase. Analyzing the data by patients, Dowling and Hirsh[97] found that in 14 (54%) both cultures were positive; in 8 (31%), the cultures with penicillinase only were positive; and in 4 (15%), the growth in the cultures without penicillinase was poor or delayed. All of the cultures studied yielded Gram-positive cocci. The cultures of four patients with endocarditis and one with bacteremia during therapy with penicillin became negative sooner in the medium without penicillinase. The authors concluded that penicillinase should be added routinely to blood cultures.

Carleton and Hamburger[98] investigated the ratios of units of penicillinase derived from *Bacillus cereus* to the micrograms of penicillin destroyed and found these ratios to range from 5:1 for penicillin itself to 60:1 for oxacillin. On the basis of peak blood levels of methicillin as high as 100 μg/ml in patients with renal failure, these investigators added an excess of *B. cereus* penicillinase to the blood cultures of two patients being treated with nafcillin for staphylococcal endocarditis and five dogs being treated with oxacillin for experimental staphylococcal endocarditis. In all cases, cultures were positive in broth containing penicillinase; none were positive in media without penicillinase. Pour plates containing penicillinase were negative in both patients, but were positive in the dogs with colony counts ranging from 4 to 73 per milliliter.

Although it is certainly advisable for blood to be taken for culture prior to the initiation of antimicrobial therapy, circumstances do occur in which this practice cannot be followed. Werner et al.[30] reported 97% positivity in 472 blood cultures obtained from 129 cases with endocarditis that had received no antimicrobial agents within 2 weeks prior to the blood culture, in contrast to 91% positivity in 209 blood cultures from 49 cases that had received antimicrobial agents within that time. This difference was statistically significant ($p < 0.02$).

Thiol medium has the ability, according to its manufacturer to inactivate penicillin, streptomycin, and sulfonamides. Indeed, Szawatkowski[59] has recently documented the inactivation by Thiol medium without SPS of many antimicrobial agents, including penicillins, cephalosporins, lincomycins, aminoglycosides, tetracycline, and chloramphenicol. This information is interesting but of uncertain significance, since parallel studies at the Mayo Clinic of TSB and Thiol, without and with SPS, have failed to demonstrate any significantly increased numbers of isolates from Thiol.[44,46] Moreover, the blood cultures from 70 patients with streptococcal endocarditis were examined to determine if positivity in Thiol and concurrent negativity or delayed positivity in TSB could be associated with recent antimicrobial therapy.[44] No such relationship could be established.

As previously noted, SPS inactivates both aminoglycosides and polymyxins.[78-80] The degree of inactivation of aminoglycosides is, however, medium-dependent.[79] Additional studies by Traub and Lowrance[79] failed to demonstrate any effects of SPS on carbenicillin, lincomycin, or amphotericin B.

The importance of the effects of residual amounts of antibiotic in the blood added to broth is not entirely clear. The blood is routinely diluted by at least 1:10, SPS is ordinarily in the medium, and subcultures should be performed early. Nonetheless, the possibility exists that a large residual concentration, particularly of penicillins and cephalosporins, remains even after dilution of the blood. In such cases, it is probably appropriate to consider adding β-lactamase to the culture. Due caution should be exercised when adding penicillinase, since its contamination in use has produced spurious results in cultures of blood. Norden[99] reported an apparent outbreak of *E. coli* bacteremias related to contamination of the penicillinase stock solution and coined the term "pseudosepticemia" to describe the problem. Faris and

Sparling[100] experienced a similar problem with spuriously positive blood cultures containing *Mima polymorpha.*

Adding penicillinase to blood culture bottles is not routine at the Mayo Clinic, unless requested by the attending physician. When requested, the enzyme is added to only one of the two or three bottles routinely inoculated with blood, and concurrently, the enzyme is sterility tested by adding a few drops of it to a tube containing thioglycollate medium supplemented with 10% serum. The bottle to which the penicillinase has been added is labeled accordingly, and a notation is also made on the culture report.

The only other additive which has been advocated by some for routine use in blood culture media is cysteine. In 1961, Frenkel and Hirsch[101] reported the isolation of certain strains of nonhemolytic streptoccocci from blood cultures of patients with bacterial endocarditis and with otitis media which appeared to require for growth some substance secreted by other bacteria, as they formed satellite colonies about colonies of other bacteria. They found that these streptococci grew well in media containing cysteine in concentrations of 0.5 to 1 mg/ml. Sodium thioglycollate, reduced glutathione, and thiomalic acid were also effective, but cystine, cysteic acid, methionine, sodium thiosulphate, and elementary sulfur were not. The authors reported that, as the content of sulfhydryl groups was decreased, the streptococci became increasingly pleomorphic with swellings, global forms, and elongations. Two of the strains grew as typical L-form colonies on osmotically stabilized media. Cayeux et al.[102] isolated streptococci from three patients with endocarditis which grew as satellite colonies about other streptococci belonging to groups A, B, C, G, and H, *S. aureus, Sarcina lutea,* and *E. coli.* Growth was stimulated by L-cysteine in low concentrations by itself and by glutathione or L-cysteine in the presence of thioglycollate or dithiothreitol.

George[103] encountered a streptococcus in 1972 which showed satellitism to *E. coli* and collected an additional six strains resembling the first. None grew on osmotically stabilized media, and all grew in cooked-meat medium and in Thiol medium. No tests were performed to determine the sulfhydryl group requirements of these strains; however, pyridoxine hydrochloride was found to support their growth. George also cited personal communications by Burdon and by Rogers, who had observed similar streptococci, the former in Thiol blood cultures from two patients with endocarditis and the latter in mixed wound infections Whether these strains were the same as those reported to require sulfhydryl groups is not known.

McCarthy and Bottone[104] isolated nine such strains from blood cultures either in Columbia broth, in which the cysteine content had been increased to 0.05%, or in other conventional media including BHI, TSB, and thioglycollate. Eight of these strains produced satellite colonies on agar cross streaked with *S. aureus.* Satellite colonies also grew around *S. epidermidis, Streptococcus faecalis, S. pyogenes, E. coli, Enterobacter, Klebsiella pneumoniae,* and *Candida albicans.* All grew promptly in subcultures to thioglycollate medium and on agar supplemented with cysteine, but not on agar supplemented with L-cystine, L-methionine, or sodium thiosulfate. Three of the strains grew only anaerobically.

The Mayo Clinic has isolated such a strain from the blood of a patient with endocarditis, and has attempted to determine whether additional isolates of these streptococci might be found if TSB were routinely supplemented with 0.05% cysteine. This did not prove to be the case in a parallel study of nearly 6000 blood cultures (Table 18).[87] It appears from the literature on this subject that these strains are usually detectable by macroscopic examination of blood culture bottles. Obviously, they would not be detected by routine subcultures unless media supplemented with L-cysteine were inoculated. They can be subcultured to thioglycollate but are generally difficult to characterize. At any rate, the finding of streptococci, which cannot be subcultured onto conventional agar media from blood culture media, cannot be casually dismissed as representing dead organisms with no clinical significance. Most would appear to fall into the viridans group of streptococci. The Clinic's case responded well to a conventional therapeutic regimen for viridans streptococcal endocarditis.

As has been illustrated, the value of these additives to blood culture media has not been clear-cut, with the notable exception of SPS. Sucrose and SAS were introduced into commercially prepared blood culture media without careful study of their potential adverse effects and, while there may be some benefit in using hypertonic media in selected cases, none has been

TABLE 18

Numbers of Isolates in Positive Cultures by Medium[a]

Organism	Both media	TSB with SPS only	TSB with SPS and cysteine only	p
Bacillus	1	7	1	NS
Clostridium	0	2	0	NS
Corynebacterium	14	35	29	NS
Escherichia	37	7	10	NS
Klebsiella	10	7	7	NS
Enterobacter	1	2	0	NS
Serratia	6	1	1	NS
Proteus	14	2	2	NS
Cardiobacterium	4	3	0	NS
Haemophilus	9	1	0	NS
Streptococcus				
S. pneumoniae	8	2	0	NS
Viridans group	24	3	4	NS
Group A	7	2	0	NS
Group D	5	2	4	NS
Alcaligenes	0	3	0	NS
Neisseria	0	1	0	NS
Bacteroidaceae	16	4	1	NS
Staphylococcus				
S. aureus	38	11	3	NS
S. epidermidis	15	20	16	NS
Veillonella	1	0	0	NS
Pseudomonas	17	2	2	NS

[a]NS, not significant.

From Washington, J. A., II, Hall, M. M., and Warren, E., *J. Clin. Microbiol.*, 1, 79, 1975. With permission.

demonstrated for SAS. One should, therefore, be cautious about adopting new additives in routine practice without critically reviewing the available scientific literature.

Incubation

Atmosphere

Until the advent of commercially prepared vacuum blood culture bottles, little attention was paid to the atmosphere in which blood cultures were incubated. Primary emphasis had been placed on the selection of media appropriate for the recovery of aerobes and anaerobes. The increasing use of vacuum blood culture bottles during the late 1960s and especially during the 1970s, however, has spawned a series of studies addressing themselves more specifically to the importance of the atmosphere of incubation in the recovery of various bacteria and fungi from cultures of the blood.

Knepper and Anthony[105] added different sizes of inocula of *P. aeruginosa, H. influenzae, N. meningitidis, S. aureus,* and *E. coli* to unvented and vented vacuum blood culture bottles containing BHI and found that, with the exception of *P. aeruginosa,* all the bacteria tested grew equally well in both bottles, even from the smallest inocula added (<10 colony-forming units). With inocula of *P. aeruginosa* ranging from <10 to 1.1×10^4 colony-forming units, no visible turbidity developed after 48 hr of incubation in the unvented bottles, despite the fact that organisms could be seen in Gram's-stained smears of the medium and could be grown in subcultures from the broth. In contrast, visible growth was noted in all instances after 48-hr incubation of the vented bottle. Of particular interest was the finding that growth of *P. aeruginosa* to a level of approximately 10^7 colonies per milliliter did occur in the unvented bottle, with or without blood; however, this growth was not detectable macroscopically by any of the usual parameters. These data confirmed those reported earlier by Slotnick and Sacks,[106] who found no macroscopic evidence of growth of various species of *Pseudomonas* in unvented bottles containing Thiol medium or TSB. In both of these studies, it was found that organisms could be subcultured from the unvented bottles for extended periods, and the introduction of air by release of the bottle's vacuum enabled sufficient additional growth to occur for its macroscopic detection.

Gantz et al.[107] inoculated vented and unvented bottles containing dextrose phosphate broth or thioglycollate medium with varying inocula of *Candida* and *B. fragilis.* With an inoculum of 1×10^2 colony-forming units of *Candida,* only 2 of 16 unvented bottles demonstrated growth macroscopically within 10 days. In contrast, all but 2 of the 16 vented bottles demonstrated growth macroscopically in 2 days. Although some colonies of *Candida* did survive for 10 days in unvented bottles, the subcultures were inconsistently positive after this time interval. All vented and unvented bottles demonstrated growth macroscopically with an initial inoculum of more than 1×10^3 colony-forming units of *B. fragilis;* however, vented bottles in over half the instances examined failed to demonstrate growth when the inoculum was one log lower. In fact, there was no evidence of growth in subcultures of vented bottles after 10 days of incubation in over a third of the instances examined at this lower inoculum size. The authors concluded that these mutually exclusive aeration requirements dictated the routine use of both vented and unvented bottles.

Braunstein and Tomasulo[108] have more recently confirmed Knepper and Anthony's results,[105] establishing the facts that *P. aeruginosa* multiplies to levels ranging from 10^7 to 10^9 colonies per milliliter in an unvented vacuum blood culture bottle and that these organisms survive and can be subcultured. They also found, as had Gantz et al.,[107] that small inocula (10 colony-forming units) of *C. albicans* did not multiply in and could not be subcultured from unvented bottles after 48 hr of incubation.[109]

On the basis of these studies, one might be tempted to conclude, as did Braunstein and Tomasulo,[108] that by 8 to 10 hr of incubation, sufficient multiplication of *P. aeruginosa* would have occurred in an unvented blood culture bottle to yield positive results in routine subcultures, and that the practice of routinely venting one of two vacuum bottles would be unnecessary except in cases of suspected fungal sepsis.[109] This conclusion has important implications, since it would enable the laboratory to employ a single-bottle blood culture system, which is certainly convenient and obviously economical. There is, however, probably no better example of the hazards of applying data derived from model blood culture systems to the clinical situation. In fact, the two do not correlate well in this instance. Blazevic et al.[110] conducted a parallel clinical study of unvented and transiently vented vacuum bottles containing Columbia broth with SPS. All bottles without macroscopic evidence of growth were routinely subcultured after 1 and 4 days of incubation. Moreover, the broths in all such bottles were routinely smeared and Gram's stained after 1, 4, and 7 days of incubation. They found that *Pseudomonas* was isolated significantly more frequently ($p < 0.01$) from the vented bottle than from its unvented counterpart (*N.B.* despite routine subcultures!). Similar results were reported by Harkness et al.[111] in a subsequent parallel clinical study of unvented and transiently vented vacuum bottles containing TSB with SPS, again despite routine subcultures on the day of the culture's inoculation and after 1 and 5 days of incubation (Table 19).

Other clinically relevant and significant differences between vented and unvented vacuum bottles were noted in these two clinical studies. Blazevic et al.[110] recovered significantly more yeasts (*Candida, Cryptococcus,* and *Torulopsis*) from the vented bottle and significantly more *B. fragilis* ($p < 0.01$) from the unvented bottle. Harkness et al.[111] recovered significantly more ($p < 0.001$) *Candida* from the vented bottle and more anaerobes (difference not satistically significant) from the unvented bottle. Otherwise, there were no statistically significant differences in recovery rates of facultatively anaerobic Gram-positive or Gram-negative bacteria between vented and unvented bottles. This study was extended by comparing vented and unvented bottles containing TSB with SPS between October 1, 1974 and October 12, 1976. Obviously, the numbers of isolates analyzed was much larger than in the study by Harkness et al.,[111] and several significant differences emerged (Table 20). Of particular importance was the finding that the unvented bottle yielded significantly more isolates of *Escherichia, Haemophilus,* Bacteroidaceae, and *Peptococcus,* while the vented bottle yielded significantly more isolates of *Klebsiella, Acinetobacter, Pseudomonas,* and *Candida.*

Tenney et al.,[39] in addition to studying the effects of volume of blood cultured, examined the role of atmosphere of incubation by comparing the recovery rates of bacteria from blood cultured in unvented and chronically vented bottles containing supplemented peptone broth with SPS. Paired comparisons of all adequately filled tubes for organisms causing sepsis failed to demonstrate any statistically significant differences between isolation rates in the two bottles, with the exception of fungi which were recovered more frequently from the vented bottle. These data suggest that the effects of venting on the rate of recovery of organisms from blood cultures could be medium-related. The previously discussed study of supplemented peptone broth reported by Painter and Isenberg[48] tends to corroborate this hypothesis. Alternatively, one might speculate that the configuration of the Vacutainer bottle somehow limits aeration of the medium or that chronically venting the bottle may be more helpful than transiently venting it. It would appear, however, from this study that there is no advantage whatsoever to be gained by not venting supplemented peptone broth in the Vacutainer bottle (50 ml).

Ellner et al.[89] made the interesting observation

TABLE 19

Numbers of Isolates in Tryptic Soy Broth: Vented (V) and Unvented (U)

Organism	No. of isolates	No. of isolates: Both U and V	V only	U only	p value
Bacillus	28	0	22	6	0.01
Clostridium	8	4	0	4	NS[a]
Corynebacterium	160	23	58	79	NS
Escherichia	173	121	22	30	NS
Salmonella	20	17	2	1	NS
Citrobacter	10	9	0	1	NS
Klebsiella	66	43	16	7	NS
Enterobacter	14	12	1	1	NS
Serratia	12	8	4	0	NS
Proteus	30	13	8	9	NS
Eikenella	1	0	0	1	NS
Haemophilus	10	6	1	3	NS
Listeria	1	0	0	1	NS
Streptococcus					
Group A	10	7	1	2	NS
Group B	13	11	0	2	NS
Group D	56	51	1	4	NS
Group F	4	2	1	1	NS
Viridans	108	84	11	13	NS
S. pneumoniae	58	50	4	4	NS
Other streptococci	23	19	1	3	NS
Eubacterium	1	0	1	0	NS
Acinetobacter	8	3	4	1	NS
Alcaligenes	5	1	2	2	NS
Moraxella	1	0	1	0	NS
Neisseria	5	0	4	1	NS
Bacteroidaceae	53	28	8	17	NS
Micrococcus	2	0	0	2	NS
Staphylococcus					
S. epidermidis	148	39	56	53	NS
S. aureus	120	85	13	22	NS
Peptostreptococcus	6	4	0	2	NS
Peptococcus	1	0	0	1	NS
Pseudomonas	67	40	20	7	<0.05
Aeromonas	3	3	0	0	NS
Candida	28	7	21	0	<0.001

[a]NS, not significant.

From Harkness, J. L., Hall, M., Ilstrup, D. M., and Washington, J. A., II, *J. Clin. Microbiol.*, 2, 296, 1975. With permission.

that the recovery of aerobic and facultatively anaerobic bacteria and especially of *E. coli* was markedly greater in a vented bottle containing isotonic Columbia broth which was incubated initially on a mechanical shaker than in a vented bottle containing the same medium which was incubated under stationary conditions. They concluded that bacteria are optimally recovered from blood with a vented bottle containing isotonic modified Columbia broth and an unvented bottle containing the same medium made hypertonic by the addition of 10% sucrose, although evidence for recommending the latter medium was certainly not clear-cut.

With the exception of supplemented peptone broth, there is general agreement among those who have investigated the effects of atmosphere on detection of bacteremia and fungemia in clinical studies that at least two bottles, one vented and one unvented, should be used routinely for blood

TABLE 20

Numbers of Isolates in Unvented and Vented Bottles Containing Tryptic Soy Broth; October 1, 1974 to October 12, 1976

	Both unvented and vented	Unvented only	Vented only	Total	*p* value
Bacillus	2	13	31	46	<0.01
Clostridium	25	10	6	41	NS
Corynebacterium	105	339	222	666	<0.001
Escherichia	527	173	131	831	<0.02
Salmonella	19	2	4	25	NS
Citrobacter	13	3	2	18	NS
Klebsiella	154	26	50	230	<0.01
Enterobacter	49	17	18	84	NS
Serratia	54	13	22	89	NS
Proteus	60	28	28	116	NS
Providencia	3	0	0	3	NS
Cardiobacterium	4	1	3	8	NS
Haemophilus	33	32	13	78	<0.01
Streptococcus					
S. pneumoniae	101	23	15	139	NS
Viridans	164	46	51	261	NS
Group A	12	5	2	19	NS
Group B	19	6	4	29	NS
Group C	3	0	0	3	NS
Group D	147	16	16	179	NS
Group F	2	1	1	4	NS
Other	51	4	4	59	NS
Listeria	0	1	0	1	NS
Lactobacillus	4	0	1	5	NS
Eubacterium	1	2	1	4	NS
Acinetobacter	11	3	12	26	<0.05
Alcaligenes	10	5	10	25	NS
Flavobacterium	0	1	5	6	NS
Moraxella	0	0	2	2	NS
Pasteurella	4	0	0	4	NS
Neisseria	2	2	5	9	NS
Bacteroidaceae	149	78	40	267	<0.001
Micrococcus	0	2	1	3	NS
Staphylococcus					
S. aureus	318	92	94	504	NS
S. epidermidis	157	165	209	531	<0.05
Peptostreptococcus	6	5	1	12	NS
Peptococcus	1	7	0	8	<0.05
Veillonella	1	1	1	3	NS
Pseudomonas	170	36	94	300	<0.001
Campylobacter	2	1	3	6	NS
Aeromonas	11	0	1	12	NS
Candida	16	3	71	90	<0.001
Cryptococcus	0	0	3	3	NS
Torulopsis	0	1	1	2	NS

Note: NS, not significant.

cultures. Although the only significant difference between unvented and vented supplemented peptone broths is in the recovery of fungi, the use of a single bottle with this medium on a routine basis would be very undesirable because of the inadequate volume of blood (5 ml) which would be cultured. Inoculation of at least one more bottle containing the same or, preferably, a different medium, e.g., TSB or BHI, would be required.

It has been generally assumed that transient venting is sufficient to ensure the growth of pseudomonads and yeasts. A recent, albeit brief, study at the Mayo Clinic, comparing transiently and chronically vented bottles containing TSB with SPS under vacuum with CO_2, has substantiated the correctness of this assumption, as there were no statistically significant differences in the isolation rates of any groups of organisms in particular, or of all organisms in general, between the two bottles. From a technical standpoint, however, transient venting is easier to cope with than is a chronic vent.

Temperature

Once inoculated, blood cultures are customarily incubated at 35 to 37°C. There have been studies performed in which bacterial survival in blood at various temperatures was examined. Wright[33] reported in 1925 that blood stored at room temperature for 5 hr showed no diminution in its bacterial content; however, if it was stored at 37°C, there was a small reduction in bacterial content during the first 5 hr and complete destruction of bacteria within 24 hr. These studies were done with blood anticoagulated by sodium citrate or heparin. Ellner and Stoessel[112] investigated the effects of temperature and anticoagulants on bacteria likely to be encountered in blood. In tubes which contained 0.05% SPS and which were stored initially at room temperature for 6 hr and subsequently at 5°C for 18 hr, all bacteria, except for *N. meningitidis* and *N. gonorrhoeae*, remained viable.

Model and Peel[113] studied the effect of prewarming blood culture media to 37°C prior to its inoculation in a model blood culture system. They inoculated approximately four colony-forming units of various species of bacteria into digest broth with 0.1% glucose and found no significant differences in rates of recovery of the bacteria from this medium at initial temperatures of 5, 20, and 37°C.

Since fungal cultures are optimally incubated at temperatures between 25 and 30°C, and since fungal sepsis cannot be uniformly or reliably suspected on clinical grounds alone, there was interest at the Clinic in determining whether the routine incubation of a vented blood culture bottle at 30°C would enhance the recovery of fungi and if this lower temperature would, in any way, impair the recovery of bacteria. A study of 2000 blood cultures compared transiently vented bottles containing TSB at 35°C with transiently vented bottles containing BHI at 30°C; both media also contained SPS. The BHI was selected for comparison since it is preferred over soybean-casein digest for fungal cultures, and since it was previously shown that bacterial recovery rates from these two media incubated at 35°C did not differ significantly (Table 8). As seen in Table 21, TSB incubated at 35°C provided significantly more isolates of *Corynebacterium, Klebsiella, Bacteroides,* and *Pseudomonas* than did BHI at 30°C. Moreover, group D streptococci and *S. aureus* were detected significantly sooner ($p < 0.01$ by the paired t test) in TSB incubated at 35°C than in BHI at 30°C. Unfortunately, no yeasts were detected in either system in over 2000 sets of blood cultures.

In conclusion, therefore, it appears that bacteria survive at room temperature for 5 or 6 hr in blood anticoagulated with sodium citrate,[33] heparin,[33] or SPS[112] and that they survive at 5°C in SPS for at least 18 hr.[112] Moreover, there appears to be no significant advantage in prewarming blood culture media before use. Finally, incubation at 30°C, rather than at 35°C, both significantly impairs the rate of recovery and prolongs the interval to detection of growth of certain groups of bacteria.

Duration

A review of the literature shows that there has been general agreement for many years to incubate blood cultures for at least 1 week and often for at least 2 weeks; however, the documentation needed to support this practice is not ample. Fox and Forrester[114] reported in 1940 that a third of the cultures which they examined (macroscopically only for the first 5 or 6 days) became positive within 24 hr after inoculation, that 90% became positive within 4 days, and that 99% had become positive within 10 days. Only 0.8% became positive after 10 to 14 days of incubation. They noted that all bacteria encountered after 10 to 14 days of incubation were streptococci from pulmonary abscesses, endocarditis, and surgical sepsis including osteomyelitis.

Ellner[40] reviewed approximately 40,000 blood cultures collected over a 10-year period and failed to find a single instance in which a culture became positive after the fifth day, despite routine incubation times of at least 14 days. In a study of blood

TABLE 21

Number of Isolates in Positive Cultures by Temperature of Incubation

Organism	Number positive at				
	Both 35 and 30°C	35°C only	30°C only	Total	*p* value
Bacillus	0	4	0	4	NS[a]
Clostridium	2	0	0	2	NS
Corynebacterium	1	12	3	16	<0.05
Escherichia	35	8	3	46	NS
Citrobacter	0	3	0	3	NS
Klebsiella	11	6	0	17	<0.05
Enterobacter	3	0	3	6	NS
Serratia	2	0	0	2	NS
Proteus	1	5	0	6	NS
Haemophilus	2	3	0	5	NS
Streptococcus					
S. pneumoniae	0	3	1	4	NS
Viridans	1	1	1	3	NS
Group B	0	1	0	1	NS
Group D	15	1	2	18	NS
Acinetobacter	0	1	1	2	NS
Pasteurella	3	0	0	3	NS
Bacteroides	4	9	1	14	<0.05
Staphylococcus					
S. aureus	22	7	2	31	NS
S. epidermidis	12	18	10	40	NS
Peptococcus	1	0	0	1	NS
Pseudomonas	12	6	0	18	<0.05
Aeromonas	2	0	0	2	NS

[a]NS, not significant.

cultures in a system of 3 sets of 4 tubes each, similar to that described by Henrichsen and Bruun,[85] Effersøe[115] randomly selected 240 sets of 12 tubes from among sets which had remained macroscopically negative during the first 7 days of incubation. One third of the 240 sets selected was examined further on the same day, another third was incubated and examined after a total of 2 weeks and the remaining third was incubated and examined after a total of 3 weeks. Growth was found in 1 set of the first group, in 2 sets of the second group, and in 3 sets of the third group of cultures. Of these delayed positive cultures, only those 3 detected after the third week of incubation were considered to have been clinically significant. It was concluded that these few additional positive cultures were not sufficient to warrant routine incubation of blood cultures for 2 or 3 weeks.

Hall et al.[46] examined 8654 sets of blood cultures in TSB and Thiol medium, both containing SPS, daily for 14 days. The cumulative percentages of positive cultures of some commonly encountered species are shown in Table 22, and it is apparent that, even disregarding presumed contaminants, not all of the isolates represented by the species listed in this table were recovered within the first 7 days of incubation and certainly not within the first 4 days. This issue has been reexamined since the advent at the Clinic of the routine early subculture,[111] and the results are shown in Figure 4. There is wide variation in the times required for the detection of some species of the Enterobacteriaceae. For example, approximately 70 and 95% of cultures positive with *E. coli* were detected within the first 24 and 48 hr, respectively, of incubation. Nearly 85% of cultures containing *K. pneumoniae* were detected within the first 24 hr, and all were detected within the first 48 hr of incubation. In contrast, only 50% of cultures with *S. marcescens* were detected within the first 24 hr, and the remainder required between 3 and 6 days of incubation for their detection.

The issue of how long to incubate blood cultures routinely is not easily resolved. In general,

TABLE 22

Cumulative Percentage Positive of Some Commonly Isolated Species, by Medium

	TSB						Thiol					
	By day (%)						By day (%)					
Organism	1	2	3	4	7	No. positive	1	2	3	4	7	No. positive
Bacillus			14	43		7					43	7
Clostridium	67	100				3	67	100				3
Corynebacterium				3	41	78	4				21	28
Escherichia	70	88	91	92	96	109	55	93	96		99	97
Klebsiella	43	79	81		96	47	48	88		90	98	40
Enterobacter	67	78			100	9	75	100				8
Proteus	58	86	93		100	14	38	85		92	100	13
Haemophilus		20	90	100		10		20		40	100	5
Streptococci												
S. pneumoniae	59	100				17	29	94	100			17
Viridans	32	59	71	74	97	34	34	69	83		94	35
Group A	29	100				7	67	100				7
Group D	72	97			100	32	61	94	97			31
Bacteroidaceae	6	39	73		94	33	8	50	77		90	26
Staphylococci												
S. aureus	26	65		71	86	65	14	45	70	73	93	56
S. epidermidis	8	31	62	69	96	26	9	18	23	32	68	22
Pseudomonas	4	44	81	93	96	27	50	100				2

From Hall, M., Warren, E., and Washington, J. A., II, *Appl. Microbiol.*, 27, 187, 1974. With permission.

7 days of incubation on a routine basis is probably adequate, providing that, in cases with persistent signs and symptoms of infection but with persistently negative blood cultures, the laboratory will be notified to continue incubation of the cultures for an additional period of time. It has been the experience at the Clinic, for example, that certain bacteria, such as *Haemophilus aphrophilus, H. paraphrophilus, H. parainfluenzae, Cardiobacterium hominis,* and *Campylobacter fetus,* which may cause infective endocarditis, are commonly isolated from blood cultures during their second week of incubation.

Although it is probably not completely satisfactory, the Clinic's solution to this issue has been to incubate and examine the cultures daily for 7 days, to reincubate them for an additional 7 days, and then to reexamine them after 14 days before discarding the cultures as negative. Unfortunately, this approach does not resolve the problems associated with the incubation space necessary for all of the bottles, but it does alleviate much of the work involved during the second week when more frequent examinations are made. This approach does not signify that there is lack of communication between the clinical and laboratory staffs but rather reflects an appreciation of the difficulties involved in the clinical staff's having to notify the laboratory to retain cultures of blood from problem cases on a regular basis. Furthermore, there have simply been too many instances when the laboratory has recovered organisms beyond the first week of incubation from cultures of patients in whom a transient febrile episode was dismissed as representing an uncomplicated bacteremic episode but which turned out to be an infection of a cardiac valve or of implanted prosthetic material. It would be a mistake for a referral center to discard blood cultures after only 1 week's observation.

Examination

Blood culture bottles should be inspected macroscopically upon their receipt in the laboratory, later in the same day, and daily thereafter for at least 7 days. Any evidence of turbidity, hemolysis, gaseousness, or colonies should prompt immediate staining and subcultures of the broth. It is not at all infrequent for one or more of these visible signs to occur after only a few hours of

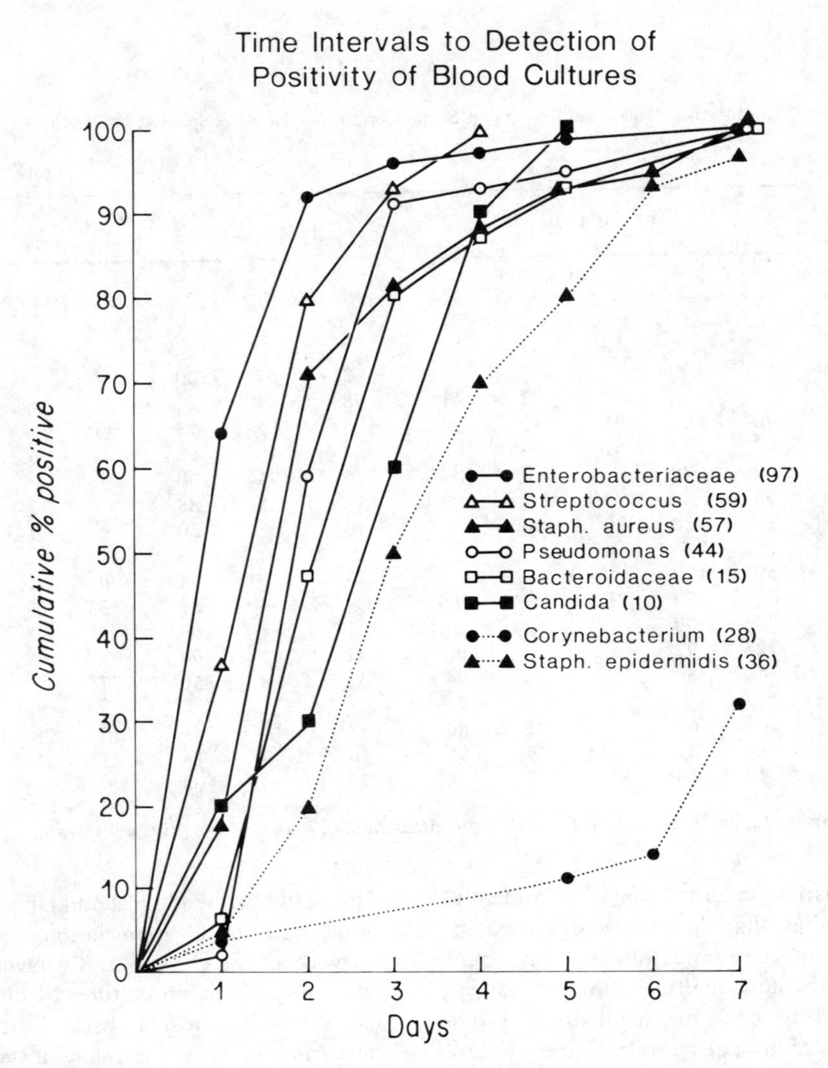

FIGURE 4. Cumulative percentage positive by day of incubation of some groups of microorganisms commonly isolated from blood.

incubation, although it is unlikely that there will be visible turbidity until there are at least 10^7 organisms per milliliter. At this point, however, a Gram's-stained smear of the medium is likely to provide a morphologic description of the organism involved. Through experience, it seems unlikely that polymicrobial bacteremia will be suspected from examination of this initial smear.

The value of routinely examining Gram's-stained smears of blood culture media which appear negative macroscopically remains controversial. Blazevic et al.[116] performed Gram's-stained smears routinely with broth from macroscopically negative bottles on the first, fourth, and seventh days of incubation. They concurrently performed routine subcultures on all but the seventh day of incubation. They initially detected 254 (23%) of 1127 positive cultures by microscopic examination (Table 23). Of the positive cultures initially detected in this manner, 49% were found on the first day, 28% on the fourth day, and 23% on the seventh day. Of 91 cultures with *Pseudomonas,* one third was initially detected microscopically, one third was detected initially by subculture, and the remaining third was initially detected macroscopically.

Hall et al.[50] routinely examined microscopically 629 blood cultures representing 1258 bottles which were macroscopically negative after their first 24 hr of incubation. In only three instances were organisms detected initially by microscopic examination, and it was concluded that routine microscopic examinations of Gram's-stained smears of blood cultures were not warranted. The differences between these two studies remain obscure. Obviously, in some people's hands routine microscopic examinations of stained blood culture media are helpful; however, in the author's opinion, a routine subculture on the day the blood is cultured is the more valuable and effective procedure (see below).

Subcultures

Positive Cultures

Subcultures of blood cultures suspected of being positive should be performed promptly and should be made onto media that are appropriate for the recovery of the types of organisms observed in the Gram's-stained smear (Table 24). If the broth is generally turbid, 2 to 4 ml are aseptically removed with a sterile needle and syringe for subcultures; however, if colonies are present in the medium, the rubber stopper is aseptically removed from the bottle, several colonies are aspirated into a Pasteur pipet for subcultures, and the stopper is replaced. Since 6 to 8% of all clinically significant bacteremias are polymicrobic,[117-120] since the initial Gram's-stained smear of a culture suspected of being positive frequently fails to provide evidence of the polymicrobial nature of the bacteremia, and since anaerobic bacteremias are often polymicrobic,[54,60] it is important to include among the media used for subculture appropriate differential media and both capneic and anaerobic environments of incubation. Von Graevenitz and Sabella[120] have, in fact, advocated the inclusion of a neomycin-blood agar plate among the media to be incubated anaerobically in order to facilitate the differentiation between facultatively anaerobic and anaerobic Gram-negative bacilli, and a Mueller-Hinton agar plate with antimicrobial discs to facilitate the detection of mixtures of Gram-negative bacilli through differential inhibitory zones. The nine discs which they placed on the Mueller-Hinton agar were ampicillin, cephalothin, chloramphenicol, colistin, nitrofurantoin, gentamicin, kanamycin, nalidixic acid, and tetracycline. They based these recommendations upon their detection of 8 to 35 cases of polymicrobial bacteremia only by use of these special subculture plates. Clinic staff attempted to see if a kanamycin-vancomycin blood agar plate incubated anaerobically would increase the detection of anaerobic Gram-negative bacilli in routine subcultures of blood cultures containing Gram-negative bacilli for a period of 2 months; any beneficial effect of this procedure could not be documented. Although not advocated per se by Von Graevenitz and Sabella,[120] direct antimicrobial susceptibility testing of positive blood cultures not only serves as an additional means of detecting mixtures of organisms, but also

TABLE 23

Method of Initial Detection of Positivity of Blood Cultures

Method	Number	%
Macroscopic	734	65
Microscopic	254	23
Subculture	139	12

Adapted from Blazevic, D. J., Stemper, J. E., and Matsen, J. M., *Appl. Microbiol.*, 27, 537, 1974.

TABLE 24

Subculture Protocol for Positive Blood Cultures

	Subculture environment				
	Capneic			Anaerobic	
Bacteria	Blood agar	EMB agar	Chocolate agar	Schaedler broth	Blood agar
Gram-positive					
Cocci	X			X	X
Bacilli	X			X	X
Gram-negative					
Cocci	X		X	X	X
Bacilli	X	X		X	X

is both practical and accurate,[121] providing that attempts are made to adjust and standardize the turbidity of the broth removed from the bottle.

Since most positive blood cultures contain only a single organism, it is usually possible to proceed directly with identification procedures and to inoculate biochemical media along with the media indicated in Table 24. Wasilauskas and Ellner[122] described a 4-hr system for the identification of bacteria wherein bottles exhibiting turbid media were opened aseptically, an aliquot of broth removed and centrifuged, and the resulting pellet (at least 0.05 ml required) utilized as a source of inoculum for a series of tests (Figure 5). Of 114 positive cultures, 112 were correctly identified in 4 hr or less. As the authors stressed, this procedure is suitable only for rapidly growing aerobic and facultatively anaerobic bacteria present in pure cultures.

With the advent of various commercially prepared kits for the identification of the Enterobacteriaceae,[123] it is not surprising that studies have now been published in which these systems have been used for the rapid identification of Gram-negative bacilli in positive blood cultures. Kocka and Morello[124] centrifuged positive blood culture media to sediment the red blood cells, removed the supernatant and centrifuged it, and resuspended the remaining pellet with 2.5 ml of distilled water in preparation for its inoculation onto the Inolex Enteric I card (Inolex Biomedical Division, Greenwood, Ill.). The reactions in the card were read after 3 hr of incubation. In addition, a Gram's-stained smear, oxidase test, and hanging-drop motility were performed. Sixty Enterobacteriaceae and one species of *Aeromonas* were correctly identified, and nine isolates of *P. aeruginosa* were recognized as being nonfermenters by the reactions obtained in the card. One organism was misidentified. With the exception of thioglycollate medium, none of the other components of the blood culture system (radiometric) which were employed exerted any adverse effects on the identification scheme. Edberg et al.[125] centrifuged the medium from a positive culture and used the pellet as the inoculum for a series of biochemical tests incorporated in reagent-impregnated strips (PathoTec®, General Diagnostics, Morris Plains, N.J.). In addition, they also performed tests for determining gelatin hydrolysis, esculin hydrolysis, bile solubility, catalase, bacitracin susceptibility, lecithinase production, motility, and serologic characteristics. Of 308 strains studied, the results of the 4 hr identification procedures agreed with those obtained by conventional procedures in 96% of instances. Blazevic et al.[126] inoculated a few

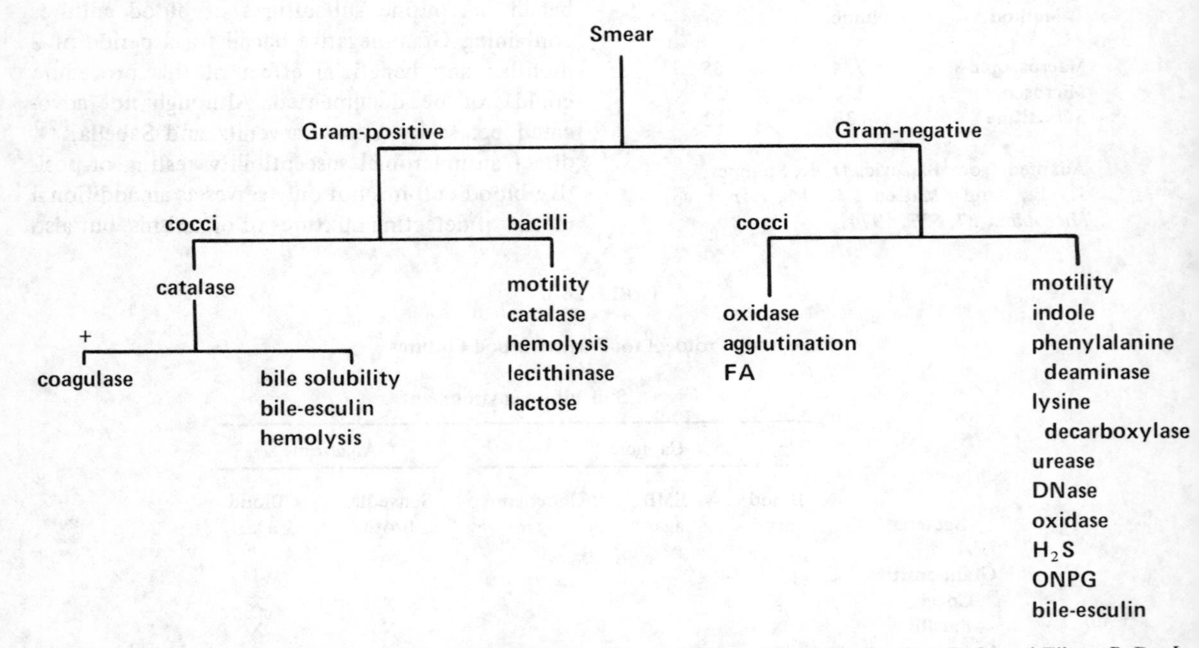

FIGURE 5. Test protocol for bacteria in positive blood cultures. (Adapted from Wasilauskas, B. L. and Ellner, P. D., *J. Infect. Dis.*, 124, 499, 1971.)

drops of blood-broth mixture containing one morphologic type of Gram-negative bacillus into 5 ml TSB which they then incubated for 4 to 6 hr until visible growth was observed and the red blood cells had settled out. They then transferred a drop of this broth to 5 ml of saline which was used to inoculate an API-20E system (Analytab Products, Inc., Plainview, N.Y.). The results with 140 cultures (of which 93 were simulated) corresponded in 99.9% of instances with the results of API-20E systems inoculated in a conventional manner with colonies from subcultures on agar plates.

The critical component of the rapid methods is a large inoculum in order to insure that a sufficient mass of the enzymes needed for the initiation and completion of certain reactions be present. The concentration by centrifugation of bacteria in the broth suspension of a positive culture apparently provides a sufficient mass of enzyme for the appropriate reactions to take place. Obviously, if one were to employ more conventional approaches to the identification of bacteria which required overnight incubation, the tests could be directly inoculated with an aliquot of the broth suspension without the necessity of preparing a centrifuged pellet. Clearly, the tests selected depend upon the Gram's-stained morphology of the organism in the blood culture; equally evident is the requirement for testing a pure culture.

Routine Subcultures

The value of routine subcultures of macroscopically negative blood cultures has been well documented and emphasized only in recent years. As discussed previously, Blazevic et al.[116] performed routine subcultures on the first and fourth day of incubation of cultures and initially detected 12% of their positive cultures in this manner (Table 23). Mayo Clinic experience shows, with cultures in unvented bottles, routine subcultures performed within 24 hr of blood collection and after 5 days of incubation provided the initial detection of 11.5% and 6.6, respectively, of positive cultures.[50] Of all of the organisms isolated by Blazevic et al.,[116] *Haemophilus influenzae, H. parainfluenzae, Moraxella,* and *N. gonorrhoeae* were initially detected only by subculture and never initially by macroscopic or microscopic examination. Of the 91 isolates of *P. aeruginosa* encountered, Blazevic et al.[116] initially detected one third macroscopically, another third microscopically, and the remaining third by subculture. Of the anaerobic bacteria isolated in their study, all but four strains of *Bacteroides* were initially detected macroscopically or microscopically. Routine anaerobic subcultures, therefore, were quite unproductive. In a study by Hall and Washington, most cultures with *P. aeruginosa* were detected in the initial subculture.[50] Potentially significant organisms recovered in the later subcultures were *P. aeruginosa* (five times), *Candida albicans* (seven times), *S. aureus* (three times), and *P. cepacia* (once). A prior study of routine anaerobic subcultures in the Clinic laboratory was also unproductive, and it would certainly seem that this type of routine subculture is not justifiable.

Routine subcultures from both the vented and unvented bottles are made on chocolate blood agar and are incubated in an atmosphere of increased CO_2 (2 to 10%) for at least 48 hr. At the Mayo Clinic, a small volume of the broth in each bottle is removed aseptically with a sterile needle and syringe and is inoculated directly onto a properly identified quadrant of a chocolate blood agar plate. In this manner, the routine subcultures of four bottles can be made on one plate. In some instances, pneumococci will not grow on chocolate blood agar, and it has been suggested that a blood agar plate be used for routine subcultures except in cases of blood cultures from the pediatric group. It is rather difficult to decide which of these media should be selected for routine subculture purposes. Although infectious diseases associated with *Haemophilus* are ordinarily limited to the pediatric population, bacteremias due to *H. influenzae, H. parainfluenzae,* and *H. paraphrophilus* are not uncommon in older patients with infective endocarditis or with impaired host-defense mechanisms. The initial detection of these species most frequently occurs in the routine subcultures on chocolate blood agar. In examining 100 consecutive cases of pneumococcal bacteremia at the Mayo Clinic, these were initially detected with rare exception within the first 24 hr of incubation either macroscopically or in the early routine subculture (see below). Although there is legitimate cause for concern about the lack of growth of pneumococci on some lots of commercially produced chocolate blood agar, the data derived from blood cultures in TSB do not indicate that there is any problem in the rapid detection of pneumococci.

The timing and frequency of routine subcultures have recently been examined by two groups of investigators. Todd and Roe[127] conducted a study in a pediatric hospital in which subcultures of Castaneda bottles containing a chocolate blood agar slant and TSB with SPS were made between 4 and 14 hr following receipt of the culture in the laboratory, and after 48 hr and 7 days of incubation. Each patient with a positive culture was reviewed to determine the clinical significance of the isolate. With the early subculture 85% of all clinically significant positive blood cultures were detected within the first 24 hr of incubation. The early subculture provided more rapid detection of *H. influenzae, E. coli, S. pneumoniae,* and other streptococci when compared to that obtained with the more conventional 24- and 48-hr stain and subculture techniques. Most of the positive cultures associated with true bacteremia were detected in the early subculture, whereas most of the cultures with contaminants were not (Table 25). Certainly, this correlation between positivity of early subcultures and clinical significance is impressive. Whether this correlation would hold up in a study of bacteremias in adults remains to be seen.

A very similar evaluation of early routine subcultures on chocolate blood agar was reported simultaneously by Harkness et al.,[111] in a study of blood cultures collected predominantly from adults. Additional routine subcultures were performed 1 and 5 days after inoculation of the culture. The number of cultures initially detected macroscopically and in each of the routine subcultures are shown in Table 26. Nearly 50% of all positive cultures were initially detected by the early routine subculture; an additional 5% and 1% were initially detected in the routine subcultures performed 1 and 5 days, respectively, after inoculation of the culture. Interestingly, 12 of 15 pneumococci were detected in the early subculture. Routine subcultures at the Clinic are currently being done on the day the cultures are inoculated and after 48 hr of incubation. All bottles received by 10:00 a.m. are subcultured at that time, and all those received by 4:00 p.m. are subcultured then. In the Clinic's centralized laboratory practice, blood culture bottles are transported to the laboratory by a courier from approximately 7:00 a.m. until 11:00 p.m. Blood cultures collected between 11:00 p.m. and 7:00 a.m. are incubated in each hospital until morning and then transferred to the central laboratory. In general, blood culture bottles require at least 1 hr between their collection and incubation in the laboratory. Early subcultures are done on a batch basis twice daily with all bottles received up to the nearest hour at which subcultures are performed. The yield from the 48-hr subculture remains small, and it seems likely that this second routine subculture could be omitted without serious consequences. No further routine subcultures are performed.

The value of the early subculture has been clearly established by these two studies. It is far less time consuming and tedious to perform an early subculture than to examine a Gram's-stained smear of each bottle on a routine basis. Moreover, since routine subcultures apparently cannot be dispensed with in radiometric systems for detecting bacteremia,[128] the early subculture eliminates, to a considerable extent, any potential temporal advantages in the detection of bactere-

TABLE 25

Relationship Between Results of Early Subculture and Clinical Significance of Isolates From Blood Cultures

Clinically significant	Early subculture	
	Positive	Negative
Yes	47	6
No	7	41

Adapted from Todd, J. K. and Roe, M. H., *Am. J. Clin. Pathol.*, 64, 694, 1975.

TABLE 26

Means of Detection of 128 Positive Blood Cultures

	Routine subculture (days)				
Cultures	<1	1	5	Macroscopically positive	Total
Number	61	7	1	59	128
Percent	48	5	1	46	100

From Harkness, J. L., Hall, M., Ilstrup, D. M., and Washington, J. A., II, *J. Clin. Microbiol.*, 2, 296, 1975. With permission.

mia which radiometric techniques may have to offer.

The problem of detecting polymicrobial bacteremias has already been discussed; however, their recognition has usually been based upon the findings in an initial subculture of a blood culture suspected of being positive. Harkness et al.[111] resubcultured 393 positive cultures representing 241 patients after 7 additional days of incubation and found 6 additional isolates which had not been identified in the original subcultures. These isolates were *S. aureus* in 4 instances, *Streptococcus anginosus* in 1, and *Klebsiella* in 1. It appears that an additional yield of 1.5% from previously positive cultures did not justify the extra effort involved for these subcultures on a routine basis.

One important aspect of routine subcultures does require study, and that is whether the use of a Castaneda (biphasic) bottle would obviate the need for routine subcultures, if one were to tilt it on the day of its inoculation and 48 hr later, so that the blood-broth mixture flows over its agar surface. Ideally, such a study would involve a comparison of two bottles under vacuum with CO_2, containing the same basal media and SPS, and the inoculation of equal volumes of blood (10 ml per bottle). Unfortunately, the major media manufacturers are reluctant to make Castaneda bottles because the agar breaks up or comes loose in shipment. The one manufacturer who has solved this problem through an ingenious bottle design (BioQuest, Cockeysville, Md.) makes a small bottle into which only 5 ml of blood can be inoculated. A new biphasic blood culture bottle design has recently been introduced into the U.S. from Australia (Medical Plastic Pty. Ltd., Adelaide); however, plans for its distribution to individual laboratories or to media manufacturers remain uncertain at this time and it remains to be carefully evaluated in clinical trials.

In summary, it is recommended that (1) routine subcultures of otherwise negative blood cultures be performed only on the day of their collection (the so-called early subculture) and 48 hr later; (2) the vented and unvented bottles be subcultured onto chocolate blood agar only; (3) routine anaerobic subcultures not be performed; and (4) any additional subcultures be done only in cases selected for reasons of persistence of negativity of cultures, in the face of clinical evidence of sepsis or intravascular infection.

In the case of cultures suspected of being positive, it is recommended that (1) media for subcultures be selected on the basis of the Gram's-stained morphology of the organism (Table 24); and (2) these subcultures routinely include a blood agar plate for incubation anaerobically.

Contamination of Blood Cultures

The risks of contaminating blood cultures posed by the microflora of the skin, inadequacy of antisepsis, and contamination of additives such as penicillinase solutions have already been discussed. There are, nonetheless, additional sources of this problem which merit attention.

Although the stoppered and sealed vacuum blood culture bottle is in common use today, some laboratories still prepare their own bottles with varying types of closures, including screw caps and cotton plugs, that must be removed in order to introduce blood into the bottle. This "open" system was shown by Conner and Mallery[129] to yield a substantially higher level of contamination than a "closed" system, wherein blood was introduced into the bottle by puncturing a rubber stopper. In fact, Conner and Mallery[129] described the first use of a blood collecting unit that consisted of 12 in. of soft rubber tubing with a needle in either end for this purpose. This unit is commonly used today for collecting blood for culture; however, collection with a syringe remains popular, and there is no evidence to suggest that one method is more effective than the other in reducing the contamination rate of the cultures. At any rate, it is recommended that a "closed" system be used routinely for blood cultures.

DuClos et al.[130] described a series of eight patients whose blood cultures were found to be contaminated with *Moraxella nonliquefaciens.* All blood for culture in their hospital was drawn into the Vacutainer tube containing supplemented peptone broth, and the source of the contaminant was traced to the nonsterile tube holders which were routinely used during the blood collection process. A protocol for cleaning these holders was implemented and eliminated this problem.

Because of the known frequency of contamination of indwelling intravascular catheters, especially after 48 hr of use, there has been serious question about the validity of culturing blood obtained through these catheters. In 174 paired comparisons between cultures of percutaneously and catheter-drawn 3-ml blood samples, Tonneson et al.[131] obtained 161 negative cultures, 6 with *Staphylococcus epidermidis* only in the catheter-drawn sample, 1 each with *Pseudomonas, Enterobacter,* and viridans streptococci in catheter-drawn samples only, and 4 with *Enterobacter* in percutaneously drawn samples only. No data are provided regarding the duration of placement of any of the catheters, the presumed significance of any of the findings, or the results of cultures of the catheters at the time of their removal. Nonetheless, the authors concluded that cultures of blood drawn through indwelling intravascular catheters were as reliable as those obtained percutaneously. Additional more critical studies of this approach seem advisable.

Spencer and Savage[132] examined the frequency and occurrence of contamination of two "open" blood culture bottles routinely subcultured after 1, 5, and 14 days of incubation on a open laboratory bench, in a safety hood, and in a laminar flow cabinet. They found that contamination rates were higher when the subcultures were performed in the safety hood and laminar flow cabinet (7 and 8.5%, respectively) than when they were carried out on a open laboratory bench (3%). They speculated that the greater turbulence of air in the hood and in the cabinet may have accounted for the somewhat higher rates of contamination of bottles subcultured in these units, but still felt that the performance of subcultures of open bottles with wire loops could be more safely carried out in the exhaust-ventilated safety hood.

Certainly, the use of a "closed" bottle system, and the performance of routine subcultures by removing an aliquot of broth through the rubber stopper with a sterile needle and syringe, should greatly reduce the risk of contamination posed by this procedure. Careful disposal of the needle and syringe should also virtually eliminate any potential hazard of the procedure to the technologist.

From time to time, there have been difficulties in manufacturing blood culture bottles which have resulted in contamination of the media with organisms such as *Bacillus.*[133] More often, however, one is apt to encounter dead microorganisms in Gram's-stained smears which are merely an annoyance. Although one might be tempted to consider them as extremely fastidious and oxygen-sensitive bacteria, they cannot be subcultured anaerobically or otherwise and can be readily discarded as being insignificant.

The distinction between contaminated and legitimately positive cultures is a problem which has been addressed by MacGregor and Beaty[134] in a study at the University of Washington. The authors reviewed all positive cultures over a 6-month period and evaluated each case for clinical evidence of bacteremia. Of 1707 cultures collected, 322 (18.9%) were positive, but 152 (47%) of these were considered to be due to contamination. Cultures were made in an "open" bottle from blood collected and transported in SPS, and pour plates were prepared during the usual daytime shift in the laboratory. Only 112 of the 233 patients (48%) had more than one blood culture inoculated, of whom 48 were considered to have been bacteremic and 64 were considered to have had their cultures contaminated. Of the 48 bacteremic patients, 33 (69%) had more than one positive culture, whereas only 7 of the 64 nonbacteremic patients (11%) had more than one positive culture. These data do substantiate the value of collecting at least two separate blood samples for culture. Pour plate cultures were positive in 68% of the bacteremic cases in which they were done and in only 29% of the nonbacteremic cases in which they were done. If those cases in which the pour plates only were positive, and a few cases in which the cultures were spuriously positive due to contamination of the anticoagulant used in the blood collection tubes were discounted, only 5.8% of contaminated broth cultures had a concurrently positive pour plate. It would appear, therefore, that pour plates may be helpful in distinguishing between legitimately positive and contaminated cultures. On the other hand, only 1 of 19 laboratories in Bartlett's survey[7] listed a percentage of positive cultures exceeding 17, suggesting that the 18.9% rate of positive cultures cited in the MacGregor and Beaty study[134] represented an unusually high rate of contamination. With lower rates of contamination, the value of pour plates may be more limited and not worth the extra effort they require, particularly if more than one set of blood cultures is inoculated per suspected bacteremic episode. Analyzing their potential sources of contamination, MacGregor and Beaty[134] did find that contamination occurred in 4% of cultures processed by the clinical microbiology laboratory, in contrast to 12.3% and 10.4% of cultures handled by the interns' and emergency night laboratories, respectively. It is amazing that, with all of the federal and state requirements for quality control and proficiency testing in clinical laboratories, house staff laboratories are permitted to continue to exist without being subjected to the same requirements.

As pointed out by MacGregor and Beaty,[134] the type of organism isolated from a blood culture is helpful in trying to determine its significance. The isolation of pneumococci, group A or B streptococci, the Enterobacteriaceae, the Bacteroidaceae, anaerobic cocci, *Haemophilus,* Neisseriae, and yeasts is nearly always clinically significant. The isolation of corynebacteria, propionibacteria, *Bacillus,* and certain groups of nonfermentative Gram-negative bacilli, such as *Alcaligenes,* is seldom clinically significant unless they are present in multiple cultures. The isolation of viridans and group D streptococci and of *Staphylococcus epidermidis* is somewhat more problematical and cannot be casually dismissed as representing contamination. Again, the presence of these organisms in multiple cultures is helpful in establishing their significance.

As has already been discussed, the continuous monitoring of presumed contamination rates on a graph, as illustrated in Figure 1, can be very helpful to the laboratory in trying to identify breaks in techniques of blood collection, problems with the blood culture bottles or the media therein, and deficiencies related to the processing of the cultures in the laboratory.

REFERENCES

1. **Evans, C. A. and Stevens, R. J.,** Differential quantitation of surface and subsurface bacteria of normal skin by the combined use of the cotton swab and the scrub methods, *J. Clin. Microbiol.,* 3, 576, 1976.
2. **Noble, W. C. and Somerville, D. A.,** *Microbiology of Human Skin,* W. B. Saunders, Philadelphia, 1974.
3. **Ahmad, F. J. and Darrell, J. H.,** Significance of the isolation of *Clostridium welchii* from routine blood cultures, *J. Clin. Pathol.,* 29, 185, 1976.
4. **Ayliffe, G. A. J. and Lowbury, E. J. L.,** Sources of gas gangrene in hospital, *Br. Med. J.,* 2, 333, 1969.
5. **Updegraff, D. M.,** A cultural method of quantitatively studying the microorganisms in the skin, *J. Invest. Dermatol.,* 43, 129, 1964.
6. **Michaud, R. N., McGrath, M. B., and Goss, W. A.,** Application of a glove-hand model for multiparameter measurements of skin-degerming activity, *J. Clin. Microbiol.,* 3, 406, 1976.
7. **Bartlett, R. C.,** Contemporary blood culture practices, in *Bacteremia: Laboratory and Clinical Aspects,* Sonnenwirth, A. C., Ed., Charles C Thomas, Springfield, Ill., 1973, 15.
8. **Washington, J. A., II,** Blood cultures: principles and techniques, *Mayo Clin. Proc.,* 50, 91, 1975.
9. **Bartlett, R. C., Ellner, P. D., and Washington, J. A., II,** *Blood Cultures,* Sherris, J. C., Ed., Cumitech 1, American Society for Microbiology, Washington, D.C., 1974.
10. **Wilson, W. R., Van Scoy, R. E., and Washington, J. A., II,** Incidence of bacteremia in adults without infection, *J. Clin. Microbiol.,* 2, 94, 1975.
11. **Kaslow, R. A., Mackel, D. C., and Mallison, G. F.,** Nosocomial pseudobacteremia: positive blood cultures due to contaminated benzalkonium antiseptic, *JAMA,* 236, 2407, 1976.
12. **Washington, J. A., II,** Blood cultures – merits of new media and techniques – significance of findings, in *Significance of Medical Microbiology in the Care of Patients,* Lorian, V., Ed., Williams & Wilkins, Baltimore, 1977, 43.
13. **Bernard, R. W., Stahl, W. M., and Chase, R. M.,** Subclavian vein catheterizations: a prospective study. II. Infectious complications, *Ann. Surg.,* 173, 191, 1971.
14. **Dunham, E. C.,** Septicemia in the new-born, *Am. J. Dis Child.,* 45, 229, 1933.
15. **Waddell, W. W., Jr., Balsey, R. E., and Grossmann, W.,** The significance of positive blood cultures in newborn infants, *J. Pediatr.,* 33, 426, 1948.
16. **Eitzman, D. V. and Smith, R. T.,** The significance of blood cultures in the newborn period, *Am. J. Dis. Child.,* 94, 601, 1957.
17. **Albers, W. H., Tyler, C. W., and Boxerbaum, B.,** Asymptomatic bacteremia in the newborn infant, *J. Pediatr.,* 69, 193, 1966.
18. **Nelson, J. D., Richardson, J., and Shelton, S.,** The significance of bacteremia with exchange transfusions, *J. Pediatr.,* 66, 291, 1965.

19. **Anagnostakis, D., Kamba, A., Petrochilou, V., Arseni, A., and Matsaniotis, N.,** Risk of infection associated with umbilical vein catheterization: a prospective study in 75 newborn infants, *J. Pediatr.,* 86, 759, 1975.
20. **Balagtas, R. C., Bell, C. E., Edwards, L. D., and Levin, S.,** Risk of local and systemic infections associated with umbilical vein catheterization: a prospective study in 86 newborn patients, *Pediatrics,* 48, 359, 1971.
21. **Chow, A. W., Leake, R. D., Yamauchi, T., Anthony, B. F., and Guze, L. B.,** The significance of anaerobes in neonatal bacteremia: analysis of 23 cases and review of the literature, *Pediatrics,* 54, 736, 1974.
22. **Thirumoorthi, M. C., Keen, B. M., and Dajani, A. S.,** Anaerobic infections in children: a prospective survey, *J. Clin. Microbiol.,* 3, 318, 1976.
23. **Cowett, R. M., Peter, G., Hakanson, D. O., and Oh, W.,** Reliability of bacterial culture of blood obtained from an umbilical artery catheter, *J. Pediatr.,* 88, 1035, 1976.
24. **Gotoff, S. P. and Behrman, R. E.,** Neonatal septicemia, *J. Pediatr.,* 76, 142, 1970.
25. **Crowley, N.,** Some bacteraemias encountered in hospital practice, *J. Clin. Pathol.,* 23, 166, 1970.
26. **Belli, J. and Waisbren, B. A.,** The number of blood cultures necessary to diagnose most cases of bacterial endocarditis, *Am. J. Med. Sci.,* 232, 284, 1956.
27. **Blount, J. G.,** Bacterial endocarditis, *Am. J. Med.,* 38, 909, 1965.
28. **Cherubin, C. E. and Neu, H. C.,** Infective endocarditis at the Presbyterian Hospital in New York City from 1938–1967, *Am. J. Med.,* 51, 83, 1971.
29. **Hayward, G. W.,** Infective endocarditis: a changing disease. I, *Br. Med. J.,* 2, 706, 1973.
30. **Werner, A. S., Cobbs, C. G., Kaye, D., and Hook, E. W.,** Studies on the bacteremia of bacterial endocarditis, *JAMA,* 202, 199, 1967.
31. **Barritt, D. W. and Gillespie, W. A.,** Subacute bacterial endocarditis, *Br. Med. J.,* 1, 1235, 1960.
32. **Cates, J. E. and Christie, R. V.,** Subacute bacterial endocarditis: a review of 442 patients treated in 14 centres appointed by the Penicillin Trials Committee of the Medical Research Council, *Q. J. Med.,* 20, 93, 1951.
33. **Wright, H. D.,** The bacteriology of subacute infective endocarditis, *J. Pathol. Bacteriol.,* 28, 541, 1925.
34. **Wilson, W. R., Dolan, C. T., Washington, J. A., II, Brown, A. L., Jr., and Ritts, R. E., Jr.,** Clinical significance of postmortem cultures, *Arch. Pathol.,* 94, 244, 1972.
35. **Minkus, R. and Moffet, H. L.,** Detection of bacteremia in children with sodium polyanethol sulfonate: a prospective clinical study, *Appl. Microbiol.,* 22, 805, 1971.
36. **Franciosi, R. A. and Favara, B. E.,** A single blood culture for confirmation of the diagnosis of neonatal septicemia, *Am. J. Clin. Pathol.,* 57, 215, 1972.
37. **Dietzman, D. E., Fischer, G. W., and Schoenknecht, F. D.,** Neonatal *Escherichia coli* septicemia – bacterial counts in blood, *J. Pediatr.,* 85, 128, 1974.
38. **Hall, M. M., Ilstrup, D. M., and Washington, J. A., II,** Effect of volume of blood cultured on detection of bacteremia, *J. Clin. Microbiol.,* 3, 643, 1976.
39. **Tenney, J. H., Reller, L. B., Stratton, C. W., and Wang, W-L. L.,** Controlled evaluation of volume of blood cultured and atmosphere of incubation in detection of bacteremia, abstr. no. 309, *16th Interscience Conf. on Antimicrobial Agents and Chemotherapy,* Chicago, October 1976.
40. **Ellner, P. D.,** System for inoculation of blood in the laboratory, *Appl. Microbiol.,* 16, 1892, 1968.
41. **Katz, L., Johnson, D. L., Neufeld, P. D., and Gupta, K. G.,** Evacuated blood-collection tubes – the backflow hazard, *Can. Med. Assoc. J.,* 113, 208, 1975.
42. **Hoffman, P. C., Arnow, P. M., Goldmann, D. A., Parrott, P. L., Stamm, W. E., and McGowan, J. E., Jr.,** False-positive blood cultures: association with nonsterile blood collection tubes, *JAMA,* 236, 2073, 1976.
43. **Washington, J. A., II,** The microbiology of evacuated blood collection tubes, *Ann. Intern. Med.,* 86, 186, 1977.
44. **Washington, J. A., II,** Comparison of two commercially available media for detection of bacteremia, *Appl. Microbiol.,* 22, 604, 1971.
45. **Washington, J. A., II,** Evaluation of two commercially available media for detection of bacteremia, *Appl. Microbiol.,* 23, 956, 1972.
46. **Hall, M., Warren, E., and Washington, J. A., II,** Detection of bacteremia with liquid media containing sodium polyanetholesulfonate, *Appl. Microbiol.,* 27, 187, 1974.
47. **Rosner, R.,** Evaluation of four blood culture systems using parallel culture methods, *Appl. Microbiol.,* 28, 245, 1974.
48. **Painter, B. G. and Isenberg, H. D.,** Clinical laboratory evaluation of the fifty-milliliter Vacutainer blood culture tube, *J. Clin. Microbiol.,* 2, 99, 1975.
49. **Morello, J. A. and Ellner, P. D.,** New medium for blood cultures, *Appl. Microbiol.,* 17, 68, 1969.
50. **Hall, M., Warren, E., and Washington, J. A., II,** Comparison of two liquid blood culture media containing sodium polyanetholesulfonate: Tryptic soy and Columbia, *Appl. Microbiol.,* 27, 699, 1974.
51. **Castaneda, M. R.,** A practical method for routine blood cultures in brucellosis, *Proc. Soc. Exp. Biol. Med.,* 64, 114, 1947.
52. **Scott, E. G.,** A practical blood culture technique, *Am. J. Clin. Pathol.,* 21, 290, 1951.
53. **Coetzee, E. F. C. and Johnson, R. S. A.,** An improved blood culture medium, *Med. Lab. Technol.,* 31, 299, 1974.
54. **Wilson, W. R., Martin, W. J., Wilkowske, C. J., and Washington, J. A., II,** Anaerobic bacteremia, *Mayo Clin. Proc.,* 47, 639, 1972.
55. **Reed, G. B. and Orr, J. H.,** Cultivation of anaerobes and oxidation-reduction potentials, *J. Bacteriol.,* 45, 309, 1943.

56. **Washington, J. A., II and Martin, W. J.,** Comparison of three blood culture media for recovery of anaerobic bacteria, *Appl. Microbiol.,* 25, 70, 1973.
57. **Shanson, D. C.,** An experimental assessment of different anaerobic blood culture methods, *J. Clin. Pathol.,* 27, 273, 1974.
58. **Forgan-Smith, W. R. and Darrell, J. H.,** A comparison of media used *in vitro* to isolate non-sporing Gram-negative anaerobes from blood, *J. Clin. Pathol.,* 27, 280, 1974.
59. **Szawatkowski, M. V.,** A comparison of three readily available types of anaerobic blood culture media, *Med. Lab. Sci.,* 33, 5, 1976.
60. **Marcoux, J. A., Zabransky, R. J., Washington, J. A., II, Wellman, W. E., and Martin, W. J.,** Bacteroides bacteremia, *Minn. Med.,* 53, 1169, 1970.
61. **Waterworth, P. M.,** The lethal effect of tryptone-soya broth, *J. Clin. Pathol.,* 25, 227, 1972.
62. **Kracke, R. R. and Teasley, H. E.,** The efficiency of blood cultures with report of a new method based on complement fixation, *J. Lab. Clin. Med.,* 16, 169, 1930.
63. **Roome, A. P. C. H. and Tozer, R. A.,** Effect of dilution on the growth of bacteria from blood cultures, *J. Clin. Pathol.,* 21, 719, 1968.
64. **Lowrance, B. L. and Traub, W. H.,** Inactivation of the bactericidal activity of human serum by Liquoid (sodium polyanetholsulfonate), *Appl. Microbiol.,* 17, 839, 1969.
65. **Von Haebler, T. and Miles, A. A.,** The action of sodium polyanethol sulphonate ("Liquoid") on blood cultures, *J. Pathol. Bacteriol.,* 46, 245, 1938.
66. **Penfold, J. B., Goldman, J., and Fairbrother, R. W.,** Blood-cultures and selection of media, *Lancet,* 1, 65, 1940.
67. **Garrod, P. R.,** The growth of *Streptococcus viridans* in sodium polyanethyl sulphonate ("Liquoid"), *J. Pathol. Bacteriol.,* 91, 621, 1966.
68. **Hoare, E. D.,** The suitability of "Liquoid" for use in blood culture media, with particular reference to anaerobic streptococci, *J. Pathol. Bacteriol.,* 48, 573, 1939.
69. **Graves, M. H., Morello, J. A., and Kocka, F. E.,** Sodium polyanethol sulfonate sensitivity of anaerobic cocci, *Appl. Microbiol.,* 27, 1131, 1974.
70. **Kocka, F. E., Arthur, E. J., and Searcy, R. L.,** Comparative effects of two sulfated polyanions used in blood culture on anaerobic cocci, *Am. J. Clin. Pathol.,* 61, 25, 1974.
71. **Rosner, R.,** A quantitative evaluation of three blood culture systems, *Am. J. Clin. Pathol.,* 57, 220, 1972.
72. **Hall, M. M., Warren, E., Ilstrup, D. M., and Washington, J. A., II,** Comparison of sodium amylosulfate and sodium polyanetholsulfonate in blood culture media, *J. Clin. Microbiol.,* 3, 212, 1976.
73. **Wilkins, T. D. and West, S. E. H.,** Medium-dependent inhibition of *Peptostreptococcus anaerobius* by sodium polyanetholsulfonate in blood culture media, *J. Clin. Microbiol.,* 3, 393, 1976.
74. **Finegold, S. M., Ziment, I., White, M. L., Winn, W. R., and Carter, W. T.,** Evaluation of polyanethol sulfonate (Liquoid) in blood cultures, *Antimicrob. Agents Chemother. – 1967,* 692, 1968.
75. **Eng, J.,** Effect of sodium polyanethol sulfonate in blood cultures, *J. Clin. Microbiol.,* 1, 119, 1975.
76. **Eng, J. and Iveland, H.,** Inhibitory effect in vitro of sodium polyanethol sulfonate on the growth of *Neisseria meningitidis, J. Clin. Microbiol.,* 1, 444, 1975.
77. **Belding, M. E. and Klebanoff, S. J.,** Effect of sodium polyanetholesulfonate on antimicrobial systems in blood, *Appl. Microbiol.,* 24, 691, 1972.
78. **Kocka, F. E., Magoc, T., and Searcy, R. L.,** Action of sulfated polyanions used in blood culture on lysozyme, complement and antibiotics, *Ann. Clin. Lab. Sci.,* 2, 470, 1972.
79. **Traub, W. H. and Lowrance, B. L.,** Media-dependent antagonism of gentamicin sulfate by Liquoid (sodium polyanetholsulfonate), *Experientia,* 25, 1184, 1969.
80. **Edberg, S. C., Bottenbley, C. J., and Gam, K.,** Use of sodium polyanethol sulfonate to selectively inhibit aminoglycoside and polymyxin antibiotics in a rapid blood level antibiotic assay, *Antimicrob. Agents Chemother.,* 9, 414, 1976.
81. **Kocka, F. E., Magoc, T., and Searcy, R. L.,** New anticoagulant for combating antibacterial activity of human blood, *Proc. Soc. Exp. Biol. Med.,* 140, 1231, 1972.
82. **Kocka, F. E., Arthur, E. J., Searcy, R. L., Smith, M., and Grodner, B.,** Clinical evaluation of sodium amylosulfate in human blood cultures, *Appl. Microbiol.,* 26, 421, 1973.
83. **Rosner, R.,** Comparison of recovery rates of various organisms from clinical hypertonic blood cultures by using various concentrations of sodium polyanethol sulfonate, *J. Clin. Microbiol.,* 1, 129, 1975.
84. **Rosner, R.,** Growth patterns of a wide spectrum of organisms encountered in clinical blood cultures using both hypertonic and isotonic media, *Am. J. Clin. Pathol.,* 65, 706, 1976.
85. **Henrichsen, J. and Bruun, B.,** An evaluation of the effects of a high concentration of sucrose in blood culture media, *Acta Pathol. Microbiol. Scand., Sect. B,* 81, 707, 1973.
86. **Sullivan, N. M., Sutter, V. L., Carter, W. T., Attebery, H. R., and Finegold, S. M.,** Bacteremia after genitourinary tract manipulation; bacteriological aspects and evaluation of various blood culture systems, *Appl. Microbiol.,* 23, 1101, 1972.
87. **Washington, J. A., II, Hall, M. M., and Warren, E.,** Evaluation of blood culture media supplemented with sucrose or with cysteine, *J. Clin. Microbiol.,* 1, 79, 1975.

88. **Chong, Y., Yi, K. N., and Lee, S. Y.,** Effects of high concentrations of sucrose in blood culture media with special reference to the cultivation of *Salmonella typhi, Yonsei Med. J.*, 16, 99, 1975.
89. **Ellner, P. D., Kiehn, T. E., Beebe, J. L., and McCarthy, L. R.,** Critical analysis of hypertonic medium and agitation in detection of bacteremia, *J. Clin. Microbiol.*, 4, 216, 1976.
90. **Simberkoff, M. S., Thomas, L., McGregor, D., Shenkein, I., and Levine, B. B.,** Inactivation of penicillins by carbohydrate solutions at alkaline pH, *N. Engl. J. Med.*, 283, 116, 1970.
91. **McGee, Z. A., Wittler, R. G., Gooder, H., and Charache, P.,** Wall-defective microbial variants: terminology and experimental design, *J. Infect. Dis.*, 123, 433, 1971.
92. **Lorian, V. and Waluschka, A.,** Blood cultures showing aberrant forms of bacteria, *Am. J. Clin. Pathol.*, 57, 406, 1972.
93. **Phair, J. P., Watanakunakorn, C., Linnemann, C., Jr., and Carleton, J.,** Attempts to isolate wall-defective microbial variants from clinical specimens, *Am. J. Clin. Pathol.*, 62, 601, 1974.
94. **Irwin, R. S., Seith, W. F., Jr., Woelk, W. K., and Enegren, B. J.,** Cell wall-deficient bacterial variant cultural surveillance: a useful laboratory aid, *Am. J. Med.*, 59, 129, 1975.
95. **Louria, D. B., Kaminski, T., Kapila, R., Tecson, F., and Smith, L.,** Study on the usefulness of hypertonic culture media, *J. Clin. Microbiol.*, 4, 208, 1976.
96. **Brogan, O.,** Isolation of L-forms by blood culture, *J. Clin. Pathol.*, 29, 934, 1976.
97. **Dowling, H. F. and Hirsh, H. L.,** The use of penicillinase in cultures of body fluids obtained from patients under treatment with penicillin, *Am. J. Med. Sci.*, 210, 756, 1945.
98. **Carleton, J. and Hamburger, M.,** Unmasking of false-negative blood cultures in patients receiving new penicillins, *JAMA*, 186, 157, 1963.
99. **Norden, C. W.,** Pseudosepticemia, *Ann. Intern. Med.*, 71, 789, 1969.
100. **Faris, H. M. and Sparling, F. F.,** *Mima polymorpha* bacteremia: false-positive cultures due to contaminated penicillinase, *JAMA*, 219, 76, 1972.
101. **Frenkel, A. and Hirsch, W.,** Spontaneous development of *L* forms of streptococci requiring secretions of other bacteria or sulphydryl compounds for normal growth, *Nature*, 191, 728, 1961.
102. **Cayeux, P., Acar, J. F., and Chabbert, Y. A.,** Bacterial persistence in streptococcal endocarditis due to thiol-requiring mutants, *J. Infect. Dis.*, 124, 247, 1971.
103. **George, R. H.,** The isolation of symbiotic streptococci, *J. Med. Microbiol.*, 7, 77, 1974.
104. **McCarthy, L. R. and Bottone, E. J.,** Bacteremia and endocarditis caused by satelliting streptococci, *Am. J. Clin. Pathol.*, 61, 585, 1974.
105. **Knepper, J. G. and Anthony, B. F.,** Diminished growth of *Pseudomonas aeruginosa* in unvented blood-culture bottles, *Lancet*, 2, 285, 1973.
106. **Slotnick, I. J. and Sacks, H. J.,** The growth of Pseudomonas in blood cultures, *Am. J. Clin. Pathol.*, 58, 723, 1972.
107. **Gantz, N. M., Swain, J. L., Medeiros, A. A., and O'Brien, T. F.,** Vacuum blood-culture bottles inhibiting growth of Candida and fostering growth of Bacteroides, *Lancet*, 2, 1174, 1974.
108. **Braunstein, H. and Tomasulo, M.,** A quantitative study of the multiplication of *Pseudomonas aeruginosa* in vented and unvented blood-culture bottles, *Am. J. Clin. Pathol.*, 66, 80, 1976.
109. **Braunstein, H. and Tomasulo, M.,** A quantitative study of the growth of *Candida albicans* in vented and unvented blood-culture bottles, *Am. J. Clin Pathol.*, 66, 87, 1976.
110. **Blazevic, D. J., Stemper, J. E., and Matsen, J. M.,** Effect of aerobic and anaerobic atmospheres on isolation of organisms from blood cultures, *J. Clin. Microbiol.*, 1, 154, 1975.
111. **Harkness, J. L., Hall, M., Ilstrup, D. M., and Washington, J. A., II,** Effects of atmosphere of incubation and of routine subcultures on detection of bacteremia in vacuum blood culture bottles, *J. Clin. Microbiol.*, 2, 296, 1975.
112. **Ellner, P. D. and Stoessel, C. J.,** The role of temperature and anticoagulant on the in vitro survival of bacteria in blood, *J. Infect. Dis.*, 116, 238, 1966.
113. **Model, D. G. and Peel, R. N.,** The effect of temperature of the culture medium on the outcome of blood culture, *J. Clin. Pathol.*, 26, 529, 1973.
114. **Fox, H. and Forrester, J. S.,** Clinical blood cultures. An analysis of over 5,000 cases, *Am. J. Clin. Pathol.*, 10, 493, 1940.
115. **Effersøe, P.,** The importance of the duration of incubation in the investigation of blood cultures, *Acta Pathol. Microbiol. Scand.*, 65, 129, 1965.
116. **Blazevic, D. J., Stemper, J. E., and Matsen, J. M.,** Comparison of macroscopic examination, routine Gram stains, and routine subcultures in the initial detection of positive blood cultures, *Appl. Microbiol.*, 27, 537, 1974.
117. **Hochstein, H. D., Kirkham, W. R., and Young, V. M.,** Recovery of more than 1 organism in septicemias, *N. Engl. J. Med.*, 273, 468, 1965.
118. **Lufkin, E. G., Silverman, M., Callaway, J. J., and Glenchur, H.,** Mixed septicemias and gastrointestinal disease, *Am. J. Dig. Dis.*, 11, 930, 1966.
119. **Hermans, P. E. and Washington, J. A., II,** Polymicrobial bacteremia, *Ann. Intern. Med.*, 73, 387, 1970.
120. **Von Graevenitz, A. and Sabella, W.,** Unmasking additional bacilli in Gram-negative rod bacteremia, *J. Med.* (Basel), 2, 185, 1971.

121. **Johnson, J. E. and Washington, J. A., II,** Comparison of direct and standardized antimicrobial susceptibility testing of positive blood cultures, *Antimicrob. Agents Chemother.*, 10, 211, 1976.
122. **Wasilauskas, B. L. and Ellner, P. D.,** Presumptive identification of bacteria from blood cultures in four hours, *J. Infect. Dis.*, 124, 499, 1971.
123. **Washington, J. A., II,** Laboratory approaches to the identification of Enterobacteriaceae, *Hum. Pathol.*, 7, 151, 1976.
124. **Kocka, F. E. and Morello, J. A.,** Rapid detection and identification of enteric bacteria from blood cultures, *J. Infect. Dis.*, 131, 456, 1975.
125. **Edberg, S. C., Novak, M., Slater, H. and Singer, J. M.,** Direct inoculation procedure for the rapid classification of bacteria from blood culture, *J. Clin. Microbiol.*, 2, 469, 1975.
126. **Blazevic, D. J., Trombley, C. M., and Lund, M. E.,** Inoculation of API-20E from positive blood cultures, *J. Clin. Microbiol.*, 4, 522, 1976.
127. **Todd, J. K. and Roe, M. H.,** Rapid detection of bacteremia by an early subculture technic, *Am. J. Clin. Pathol.*, 64, 694, 1975.
128. **Caslow, M., Ellner, P. D., and Kiehn, T. E.,** Comparison of the BACTEC system with blind subculture for the detection of bacteremia, *Appl. Microbiol.*, 28, 435, 1974.
129. **Conner, V. and Mallery, O. T.,** Blood culture: a clinical laboratory study of two methods, *Am. J. Clin. Pathol.*, 21, 785, 1951.
130. **DuClos, T. W., Hodges, G. R., and Killian, J. E.,** Bacterial contamination of blood-drawing equipment: a cause of false-positive blood cultures, *Am. J. Med. Sci.*, 266, 459, 1974.
131. **Tonnesen, A., Peuler, M., and Lockwood, W. R.,** Cultures of blood drawn by catheters vs. venipuncture, *JAMA*, 235, 1877, 1976.
132. **Spencer, R. C. and Savage, M. A.,** Laboratory contamination of blood cultures, *J. Clin. Pathol.*, 28, 980, 1975.
133. **Kagan, R. L., Kasten, B. L., Brenner, V. J., and MacLowry, J. D.,** Contamination of Difco Bacto blood culture bottles, *N. Engl. J. Med.*, 290, 1024, 1974.
134. **MacGregor, R. R. and Beaty, H. N.,** Evaluation of positive blood cultures: guidelines for early differentiation of contaminated from valid positive cultures, *Arch. Intern. Med.*, 130, 84, 1972.
135. **Lane, A.,** personal communication, Difco Laboratories, Detroit, 1976.
136. **Mangels, J. I., Lindberg, L. H., and Vosti, K. L.,** Comparative evaluation of three different commercial blood culture media for recovery of anaerobic organisms, *J. Clin. Microbiol.*, 5, 505, 1977.

Chapter 3

CULTURES FOR MISCELLANEOUS ORGANISMS

John A. Washington II

BRUCELLA

Although the number of reported cases of brucellosis in the U.S. has declined steadily during the last 25 years, in the past decade a rate of 0.09 to 0.15/100,000 population has been reported annually to the Center for Disease Control in Atlanta.[1] The infrequency of the disease and its protean manifestations make its diagnosis difficult. Although the presence of brucella agglutinins in serum may be helpful in approaching the diagnosis of brucellosis, a definite diagnosis cannot be established on this basis alone. The isolation of the organism from blood, another body fluid, or tissue does provide the definitive diagnosis.

Most of the principles already presented and discussed for the isolation of bacteria other than brucellae from the blood are also germane to the isolation of brucellae. As pointed out by Spink,[2] the order of magnitude of the bacteremia is low and often intermittent in nature. He has, accordingly, suggested culture of 30 to 40 ml of blood on each of several successive days. This point was corroborated by Killough et al.,[3] who found that only 70 of 100 bacteriologically proven cases of brucellosis would have been positive had only a single blood culture been made.

Castaneda et al.[4] studied the influence of medium on the isolation of brucellae from blood. They recovered the organism from 21 of 46 (45.6%) blood cultures in liver infusion broth, from 67 of 85 (78.8%) in 4% citrate solution, and from 66 of 81 (81.4%) in tryptose broth (Difco Laboratories, Detroit, Mich.). Castaneda[5] subsequently described a biphasic medium system for routine blood cultures in brucellosis wherein 15 ml of tryptose agar were poured into 100-ml flat-sided, rectangular bottles and allowed to harden on one sidewall. Ten millilitres of tryptose broth were then added; the air in each bottle was replaced with suitable mixtures of CO_2 and air, and the bottle was capped with a rubber stopper.

McCullough[6] later reported that Trypticase soy broth (BioQuest, Cockeysville, Md.) more nearly met the growth requirements of many strains of *Brucella* and recommended an atmosphere of 10% CO_2 for culture. Spink[2] found that the most satisfactory media in quantitative studies carried out at the University of Minnesota were Trypticase soy broth and agar and Brucella broth and agar (then manufactured by Albimi Laboratories, Brooklyn, N.Y. and now manufactured by several commercial sources of media). In a report of the diagnostic criteria for human brucellosis, Spink et al.[7] recommended that either of these media be used in liquid form for broth cultures or in a biphasic system according to the Castaneda[5] technique.

Broth cultures, according to Spink et al.,[7] should be subcultured on the 4th day and at regular intervals thereafter onto Trypticase soy or Brucella agar plates and should be kept for a month before being discarded as negative. Alternatively, a Castaneda bottle can be used and should be tipped at 48-hr intervals to allow the blood-broth mixture to flow over the exposed agar surface.

McCullough[6] mentioned that in chronic brucellosis, in which bacteremia was often absent and was certainly not constant, culture of arterial blood 20 min following the subcutaneous injection of 0.5 to 1.0 ml of epinephrine occasionally yielded the organism. Dalrymple-Champneys[8] described similar occasional success in obtaining a positive blood culture following the injection of typhoid vaccine.

The reported degree of success in culturing *Brucella* from patients with presumed brucellosis has varied widely. Spink[2] reported that brucellae were isolated from the blood of 139 of 244 cases (57%), while Dalrymple-Champneys[8] reported that blood cultures were successful in only 71 of the 439 cases (16%) studied. In contrast, Castaneda et al.,[4] reported that 84% of their blood cultures were positive. In a report of experiences with 224 patients with brucellosis at the Mayo Clinic from 1940 through 1958, Schirger et al.[9] described 47 patients (20%) with positive blood cultures on one or more occasions. Obviously, the rate of positivity of blood cultures in this disease is related to its type or stage.

Spink[2] has advocated that cultures of aspirates of the bone marrow be performed. In 59 patients with simultaneous blood and bone marrow cultures, there were five cases in which the marrow cultures only were positive. The converse occurred in no cases. Hamilton[10] discovered that the rates of recovery of *Brucella* from single cultures of blood and bone marrow were identical (40%), but that the organism was recovered from 86% of cases when multiple blood cultures were made.

In summary, it is necessary to perform multiple cultures of an adequate volume (30 to 40 ml per culture set) of blood to document the presence of brucellae in blood in as many cases as possible. Blood should be inoculated into Trypticase soy or Brucella medium according to the Castaneda technique and incubated for 1 month in an atmosphere containing 10% CO_2. The effect of sodium polyanetholsulfonate (SPS) on brucellae in clinical blood cultures has not been evaluated. Media for brucella blood cultures at the Mayo Clinic do not contain SPS and are also used for the cultivation of pathogenic neisseriae from blood.

LEPTOSPIRA

The number of reported cases of leptospirosis in the past decade has ranged from 41 to 93.[1] Because of the great variability in the severity of the disease, a large number of inapparent infections do occur and are detected by serologic surveys of high risk groups.

Although leptospires may sometimes be seen in blood films by darkfield or fluorescent microscopy or in silver-stained smears, their visualization is extremely difficult except during the first 8 days of illness when their numbers exceed 100,000/ml.[11] More often than not, cultures are performed by inoculating the blood into artificial media or into animals (guinea pigs, hamsters, or gerbils). While the intraperitoneal injection of from one to ten leptospires is sufficient to infect weanling hamsters,[11] most clinical laboratories are simply not equipped to handle animal inoculations and must rely upon in vitro techniques to isolate the organism.

Many different kinds of media have been described for the cultivation of leptospires. Sterile rabbit serum is ordinarily added to these media but should not exceed 14% v/v.[12] Fletcher's semisolid and Stuart's liquid media will support the growth of most leptospires and are available commercially. From one to three drops of fresh blood or blood anticoagulated with sodium oxalate or SPS should be inoculated into 5 ml of medium which is incubated at 28 or 29°C for 5 or 6 weeks.[13] Growth is greatly diminished by incubation temperatures exceeding 29°C.[12] An amount of blood in excess of this volume may be inhibitory to leptospires. Cultures are examined microscopically by darkfield illumination on a weekly basis by transferring a loopful of medium onto a glass slide and scanning it with either a low power or high dry objective.[13,14]

Leptospires are seldom, if ever, isolated from the blood beyond the acute, septicemic stage of the illness which lasts from 4 to 7 days. Most patients in whom the diagnosis is suspected on clinical grounds are already beyond this stage so that blood cultures represent a considerable waste of time and effort. Requests for leptospiral blood cultures should, therefore, be screened before actually proceeding with the cultures.

Christmas et al.[15] reported the isolation of leptospires by blood culture from 75% of acutely ill human cases in New Zealand. Six bottles of culture media were used for each patient; four bottles containing approximately 7 ml of modified Vervoort-Schuffner medium[16] and two containing 5 ml of Tween 80-albumin medium.[17] This study demonstrated the clear superiority of the latter medium over the former, but it is not known how either compares with Fletcher's semisolid medium.

Interestingly, Thorsteinsson et al.[18] reported the isolation of leptospires from ordinary media (variety not specified) inoculated with blood from a patient 2 days after the acute onset of fever, headache, and malaise. Darkfield examination, after Gram's stained smears and subcultures were negative, of a turbid layer in the broth showed leptospires. This experience suggests that such cultures be carefully scrutinized by darkfield or phase-contrast microscopy for the presence of spirochetes.

FUNGI

As stated by Krick and Remington[19] in a recent review of opportunistic invasive fungal infections in patients with leukemia and lymphoma, the unreliability of positive cultures in diagnosing opportunistic fungal infection is well known, as is the unreliability of ante-mortem or post-mortem negative cultures in excluding the diagnosis of aspergillosis and phycomycosis. Less well known perhaps is the virtual worthlessness of negative blood cultures in ruling out disseminated

TABLE 1

Positivity of Blood Cultures in *Candida* Endocarditis Following Cardiac Surgery

Interval between surgery and culture (month)	Cultures (no.)		
	Positive	Negative	Total
≤1	22	24	46
1–4	28	17	45
>4	13	7	20
?	10	1	11

Adapted from Seelig, M. S., Speth, C. P., Kozinn, P. J., Taschdjian, C. L., Toni, E. F., and Goldberg, P., *Prog. Cardiovasc. Dis.*, 17, 125, 1974.

candidiasis. This rather grim commentary about the problems involved is establishing the diagnosis of fungal sepsis is certainly not limited to patients with leukemia and lymphoma, although the occurrence of fungemia in any immunologically compromised host must always suggest the possibility of invasive disease.

Negative antemortem blood cultures have been reported in 62 to 100% of acute leukemics with autopsy-proven deep-seated candidiasis.[19] In an analysis of 91 cases of *Candida* endocarditis following cardiac surgery, Seelig et al.[20] found that candidemia was only slightly more reliable than fever as an index of this disease. Slightly fewer than half of those patients whose blood was cultured within a month after surgery had positive blood cultures, but nearly two thirds of those cultured in the next 3 months were positive (Table 1). In a more general review of published reports of postoperative fungal endocarditis, Harford[21] found that fungemia was absent in 50% of the cultures collected.

It is apparent that the rate of positivity of blood cultures in patients with fungal disease is closely related to the frequency of infection due to the filamentous fungi. Blood cultures from patients with invasive aspergillosis or with *Aspergillus* endocarditis, for example, are rarely positive.[19,21-24]

Young et al.[25] studied 70 patients from the National Cancer Institute, Bethesda, Md., who had positive fungal blood cultures ante-mortem. Fungal infection was proven on post-mortem examination in 34 (49%). In 10 (14%), the diagnosis was based upon clinical evidence and response to amphotericin B therapy; in 8 (11%), the fungemia was transient and was associated with i.v. catheters; in 14 (20%), there were unexplained saprophytic fungi isolated; and 4 (6%), unexplained isolations of *Candida* and *Cryptococcus* occurred. Those patients with disseminated mycoses had more positive blood cultures on the average than those without evidence of infection. Also, 79% of those with disseminated disease had more than one positive culture in contrast to only 31% of those without invasive fungal disease, the latter of whom had blood for cultures withdrawn through contaminated i.v. catheters. The isolation of saprophytic fungi, including *Penicillium, Cladosporium,* and *Aspergillus,* from blood cultures in this study was not associated with an infectious process and was presumed to have resulted from contamination of the culture. As previously mentioned, blood cultures from patients with infectious processes due to *Aspergillus* are rarely positive. While these guidelines may be helpful in the interpretation of positive fungal blood cultures, the frequency of negativity of fungal blood cultures in the face of disseminated mycoses certainly bears reemphasis.

The problem associated with the diagnosis of fungemia are quite complex and relate both to the nature of disseminated or intravascular fungal diseases and to the laboratory methods employed in their detection. Akbarian et al.[26] successfully produced experimental *Histoplasma* endocarditis in five dogs. Cultures were performed four to five times weekly by inoculating 2 ml of venous blood into a pour plate with blood agar base that was kept sealed at 37°C for at least 30 days and examined twice weekly. The earliest positive culture occurred 6 days after infection in one dog, and the latest occurred 27 days after infection in one dog. Two dogs had no positive cultures out of 35 performed; three dogs had 3, 10, and 5 positive cultures out of 51, 21, and 43 cultures, respectively. The authors also reviewed 15 previously reported human cases of *Histoplasma* endocarditis and found that all of the blood cultures had been negative; however, in only one case were fungi sought.

Stone et al.[27] have ascribed the low rates of recovery of *Candida* from venous blood cultures to the rapid clearance of the organism from the blood by the liver. By injecting *C. albicans* at concentrations of 10^2 to 10^6 colony-forming units per milliliter into the portal vein of dogs and culturing blood from various sites, they demonstrated that the liver was able to clear the organism at a ratio of 10,000 to 1. They recommended that blood for culture of yeasts be obtained from a peripheral artery or from the right atrium to minimize the influence of "tissue filtration" on their recovery. Obviously, the use of this procedure is severely restricted, if not precluded, in patients with thrombocytopenia secondary to their underlying disease or to its treatment. Nonetheless, this approach appears to merit further investigation in patients able to tolerate the procedure safely. More recently, Kobza et al.[28] have reported that the diagnosis of candidemia was established in a high number of cases with disseminated candidiasis by culturing blood taken from a central venous pressure catheter. All of the patients were apparently already receiving fluids and antibiotics through these catheters. While their report is of interest, authors do not specify how long these catheters had been in place, what cultures of peripheral venous blood showed when performed, or how they established the diagnosis of disseminated candidiasis in the absence of antemortem or post-mortem histological evidence. As has been pointed out in numerous reports, the mere isolation of *Candida* from patients with i.v. catheters is not necessarily indicative of disseminated candidiasis.

Literature on fungemia generally lacks details of the methods used for fungal culture. It seems safe to assume, therefore, that the techniques employed varied widely and in many instances they were those used for conventional bacterial blood cultures. Although, as will be discussed below, properly vented vacuum blood culture bottles provide a satisfactory means of isolating the yeast-like fungi (including *Candida, Cryptococcus,* and *Torulopsis*), they are not particularly suitable for isolating the filamentous fungi and are unlikely to be incubated for a period of time sufficient for these fungi to grow.

Fungemia has not been characterized adequately to make any specific recommendations regarding the number of cultures that should be collected or the volume of blood that should be cultured. It is obvious from the studies by Akbarian et al.[26] that, in experimental *Histoplasma* endocarditis, few blood cultures were positive, and it seems reasonable on the basis of clinical studies to suggest that multiple blood cultures be obtained over the course of several days when fungemia is suspected. It is difficult, however, to offer any other firm guidelines regarding the number and timing of cultures to be collected because of the absence of data on the subject and because it seems apparent that there are considerable differences between the fungemias caused by the yeast-like fungi and those caused by the filamentous fungi.

Our own practice with regard to fungal blood cultures is that (1) they must be specifically requested and (2) at least two separate collections of blood will be made within a 24-hr interval.

The yeast-like fungi may be recovered readily from transiently vented conventional vacuum blood culture bottles. Blazevic et al.[29] compared transiently vented and unvented blood culture bottles containing Columbia broth with SPS under vacuum with CO_2. *Candida, Cryptococcus,* and *Torulopsis* were recovered significantly more frequently from the vented than from the unvented bottles (Table 2). These data corroborated findings with model blood culture systems by Gantz et al.,[30] demonstrating that an inoculum of 1×10^2 colony-forming units of *Candida* produced visible turbidity in all but 2 of 16 vented bottles within 2 days and in only 2 of 16 unvented blood culture bottles by the 10th day of incubation. Roberts et al.[31] also found that in a model blood culture system using unvented vacuum bottles containing soybean-casein digest (TSB) detection times of yeasts were markedly delayed or even absent. In comparing yeast isolates from clinical blood cultures there were also significantly ($p < 0.001$) more recovered from transiently vented than from unvented blood culture bottles (Table 3). The results in transiently vented bottles containing TSB were compared with those in the conventional fungal blood culture bottle, a vented biphasic brain heart infusion (BHI) bottle, without finding any significant differences in their rates of positivity (Table 4) but with a significant difference in the time intervals to detection of positivity (Table 5). In a subsequent study comparing transiently vented and unvented vacuum blood culture bottles containing TSB, Harkness et al.[32] found significantly more ($p < 0.001$) isolates of *Candida* in the former than in the latter bottle (see Table 19, Chapter 2).

It has never been clearly proven that transiently venting vacuum blood culture bottles would be sufficient for the growth of fungi, and it has been the practice at the Mayo Clinic to place a chronic

TABLE 2

Number of Yeast-like Fungi Isolated from Vented and Unvented Vacuum Blood Culture Bottles

		No. of isolates			
Organism	Isolates	Both vented and unvented	Vented only	Unvented only	p Value
Candida	28	6	22	0	<0.01
Cryptococcus	6	0	6	0	<0.05
Torulopsis	4	0	4	0	NS[a]

[a]NS = not significant.

Adapted from Blazevic, D. J., Stemper, J. E., and Matsen, J. M., *J. Clin. Microbiol.*, 1, 154, 1975.

TABLE 3
Yeast Isolates From Blood by Medium

	TSB, vented	
TSB, unvented	Positive	Negative
Positive	7	0
Negative	20	

From Roberts, G. D., Horstmeier, C., Hall, M., and Washington, J. A., II, *J. Clin. Microbiol.*, 2, 18, 1975. With permission.

TABLE 4
Yeast Isolates from Blood by Medium

	Biphasic BHI, vented	
TSB, vented	Positive	Negative
Positive	10	8
Negative	13	

From Roberts, G. D., Horstmeier, C., Hall, M., and Washington, J. A., II, *J. Clin. Microbiol.*, 2, 18, 1975. With permission.

TABLE 5
Time Intervals to Detection of Yeasts in Matched Pairs of Positive Blood Cultures

Medium	Average days to positivity	p Value
Biphasic BHI, vented	2.6	<0.01
TSB, vented	5.2	

vent into fungal blood culture bottles. Recently, however, studies at the Clinic have shown that there were no statistically significant differences in isolation rates of yeast-like fungi between transiently and chronically vented vacuum blood culture bottles containing TSB in a parallel study of 2000 sets of cultures.

Based on findings in a model blood culture system that were essentially the same as those previously reported by Gantz et al.[30] and by Roberts et al.,[31] Braunstein and Tomasulo[33] recommended the routine use of a single vacuum bottle, which would be transiently vented following a blind subculture for anaerobes on the 2nd day to enhance the recovery of yeasts. As attractive as this recommendation sounds, it has never been proven to be as effective in a clinical trial as the immediately vented vacuum bottle for detecting yeasts. Moreover, as was discussed above, there are legitimate objections to a single bottle system because of the limited volume of blood cultured and because of the mutually exclusive atmospheric requirements of yeasts and *Pseudomonas*, on one hand, and anaerobic bacteria, on the other. It is important to reemphasize that application of results obtained in a model blood culture system to clinical blood cultures is fraught with considerable hazard.

There have been essentially no studies comparing the efficacy of various media in the recovery of fungi from blood. Many laboratories rely upon conventional bacteriological blood culture media for the isolation of fungi, whereas others inoculate 1 to 3 ml into tubes containing either BHI or Sabouraud's dextrose agar. BHI

TABLE 6

Pathogenic Fungi Detected at the Mayo Clinic in Fungal Blood Cultures: January, 1972 to June, 1974

Organism	Number of isolations
Candida albicans	74
C. parapsilosis	20
C. tropicalis	16
Cryptococcus neoformans	12
C. laurentii	2
Torulopsis glabrata	18
Torulopsis sp.	1
Histoplasma capsulatum	16

Adapted from Roberts, G. D. and Washington, J. A., II, *J. Clin. Microbiol.*, 1, 309, 1975.

according to the Castaneda technique[5] has been used at the Mayo Clinic, a choice based more on historical precedent than on scientific fact. Each bottle contains a slant of 50 ml of BHI agar and 60 ml of BHI broth and is inoculated with 10 ml of blood.[34] Despite the absence of SPS or other anticoagulant in the medium, clotting does not occur in this system. The bottles are chronically vented during their incubation at 30°C for 30 days, inspected daily for macroscopic evidence of growth, and then gently mixed so that the blood-broth mixture flows over the agar slant. The pathogenic fungi isolated from 7425 cultures during 2½-year period are listed in Table 6.

The potential value of hypertonic media in the recovery of fungi from blood cultures was studied by Roberts et al.[35] who compared unsupplemented BHI broth with BHI broth containing 15% sucrose in parallel blood cultures performed according to the Castaneda technique. Fungi were recovered more frequently ($p < 0.01$) in the absence of sucrose (Table 7).

It is apparent from this brief discussion concerning the media used for fungal blood cultures that there is little known about the subject and that more studies are needed. The principal difficulty is, however, that the ante-mortem detection of fungemia by whatever means is a relatively rare event so that meaningful data are difficult to obtain. In the study previously cited by Roberts et al.,[35] a total of 51 isolates were recovered from 3902 fungal blood cultures over an 11-month period. Moreover, all but seven of these isolates were yeast-like fungi which can be readily recovered in vented vacuum bottles used for bacterial cultures of blood.

At the Mayo Clinic, fungal blood cultures are incubated at 30°C which is suitable for the growth of the yeast-like fungi and the filamentous fungi. Parallel studies of cultures incubated at 30 and

TABLE 7

Numbers of Isolates in Blood Cultures by Medium

Organism	Both media	BHI only	BHI with sucrose only	Total positive	*P*
Candida albicans	5	9	1	15	<0.05
Candida parapsilosis	4	2	0	6	NS[a]
Candida tropicalis	3	10	5	18	NS
Cryptococcus neoformans	3	0	0	3	NS
Histoplasma capsulatum	5	1	1	7	NS
Torulopsis glabrata	0	2	0	2	NS
Total	20	24	7	51	<0.01

[a]NS, Not statistically significant.

From Roberts, G. D., Horstmeier, C. D., and Ilstrup, D. M., *J. Clin. Microbiol.*, 4, 110, 1976. With permission.

35°C have not been performed, and it is not known which of these two temperatures might yield the greater number of isolates from blood cultures. However, the isolation of filamentous fungi is facilitated at the lower temperature, as is the identification of the dimorphic fungi.

Other approaches to the detection of fungemia have been used. Blastospores of *Histoplasma capsulatum* were noted within neutrophils in a peripheral blood smear by Parsons and Zarafonetis.[36] *Candida* blastospores and pseudohyphae have been observed in blood smears by Portnoy et al.,[37] Anderson and Yardley,[38] and Silverman et al.[39] In the cases with candidiasis, the positive smears were made from peripheral blood in two instances and from blood taken from central venous pressure catheters in three instances. Although this procedure can be helpful in the rapid diagnosis of some septicemias, one must bear in mind the warning made by Humphrey[40] in 1944 to the effect that it can in no way be substituted for a blood culture and can serve merely as an adjuvant in those instances in which the organisms are relatively numerous because, in the majority of septicemias, the cultures would be positive and the smears negative.

Komorowski and Farmer[41] described a lysis-filtration blood culture technique that was compared to a conventional method employing TSB with 0.1% agar in a flask incubated at 35°C for 30 days in cases of suspected candidemia and subcultured terminally if still macroscopically negative. Blood was lysed with alkyl phenoxypolyethoxy ethanol (Triton X-100) and sodium carbonate and filtered through a membrane with a 0.45-μm pore size that was then placed onto the surface of antibiotic-free Thayer-Martin agar and incubated at 35°C in 5% CO_2 for 3 days. Of 29 isolates of *Candida* recovered by the lysis-filtration technique, only 17 were recovered in the TSB with agar. Moreover, whereas isolations in the lysis-filtration method were usually made in 24 hr and always in 48 hr, it almost invariably took more than a week for growth to become detectable in the TSB. As the authors pointed out in their article, surface growth of *Candida* was more rapid than submerged growth in broth. What role the agar added to the TSB may have played here is unknown, but it seems likely that it would have maintained the E_h of this medium at a somewhat lower level than had it not been present at all. Sullivan et al.[42] also found the lysis-filtration method to be superior to broth cultures for the recovery of yeasts from blood; however, none of the blood culture bottles used were vented. Unfortunately, the lysis-filtration technique is rather cumbersome to perform unless one can use a disposable filtration unit[43] which is not yet available commercially.

Dorn et al.[44,45] have described a blood culture technique based on centrifugation of lysed blood into a high density layer or gel that in clinical trials yielded substantially more isolates of fungi than did vented and unvented bottles containing TSB with SPS. Of 1000 cultures performed, 20 fungal isolates were obtained by the centrifugation technique, whereas only 1 was obtained in TSB. Although the contamination rates with this technique substantially exceeded those in the conventional bottles, this new approach certainly bears further study, pending its commercial development and the ease of its use.

In conclusion, there are multiple problems associated with the detection of fungemia antemortem and with the interpretation of blood cultures yielding fungi. Although the isolation rates of yeast-like fungi from conventional bacteriologic blood cultures, in vented bottles, appear to be equivalent to those obtained with media more commonly used for the isolation of fungi, this probably is not true of the filamentous fungi. Moreover, the detection of the yeast-like fungi in bacteriologic media, e.g., TSB, appears in our experience to be significantly slower than in a Castaneda bottle containing BHI broth and agar. Finally, bacteriologic media are not apt to be incubated for a period long enough for the recovery of fungi. It is, therefore, desirable to have a fungal blood culture system available for use in cases of suspected fungal sepsis. Ideally, one system would cover both bacteria and fungi; however, such a system does not appear to exist yet. It was hoped at one time that a vented Castaneda bottle containing BHI and incubated at 30°C for 30 days would not only serve for fungal cultures, but could also replace the vented TSB bottle now used as part of a bacteriologic culture set. This hope was, however, considerably dampened because of the significantly lower frequency of isolation of *Klebsiella, Bacteroides,* and *Pseudomonas* specifically, and of many organisms in general from BHI incubated at 30°C than from BHI incubated at 35°C (see Table 21, Chapter 2).

Physicians should be encouraged to request that

fungal blood cultures be performed in cases of suspected fungal sepsis. The challenges are that fungal sepsis and bacterial sepsis are often indistinguishable clinically, and that fungemia is difficult to detect by laboratory methods.

MYCOPLASMA

There are many reports in the literature documenting the isolation of mycoplasmas, usually of genital origin, from blood. Carlson et al.[46] reported the isolation of mycoplasmas from blood cultures of three children in 1951. One had an indolent ulcer of the finger, a petechial rash, leukocytosis, abdominal pain, bloody diarrhea, and a low-grade fever. The second had migratory arthritis, petechial rash, abdominal pain, and blood in his stool, and was subsequently found to have intussusception that was removed surgically. There were nine positive blood cultures in this case over a 6-week course. The third had congenital heart disease and developed a brain abscess which ruptured. Blood cultures in BHI broth from all three cases and from cerebrospinal fluid in the third case demonstrated slight cloudiness and fine granular sediment. Gram's stained smears of the sediment were negative, but methylene blue stained smears showed pleomorphic rounded structures that could be subcultured onto tryptose agar containing dextrose, peptone, and ascitic fluid, horse serum, or blood. Minute colonies were observed extending into the agar. An additional 200 blood cultures were examined, and 50 had smears with structures suggestive of mycoplasmas; however, none of the subcultures yielded any growth.

Most of the other reports of mycoplasmemia have been in patients with postpartum fever or following gynecological surgery or instrumentation. Singerland and Morgan[47] reported nine positive blood cultures over a 24-hr interval in a postpartum patient. Gram's stained smears revealed pleomorphic coccoid and filamentous bodies that grew anaerobically as minute colonies on blood agar and aerobically in horse serum broth. Stokes[48,49] reported two cases, classified for the first time as Type 1 genital strains, from liquid blood culture media containing SPS and incubated in air, 5 to 10% CO_2, and anaerobically.

Tully et al.[50] isolated *M. hominis* Type 1 from a patient who was receiving steroid therapy for nephrosis and who became septic 3 days following a therapeutic abortion. The organism was isolated from subcultures onto blood agar after 11 days of incubation of several different kinds of conventional blood culture media. The colonies were minute and grew down into the agar. The patient responded clinically to chloramphenicol, to which the organism was found to be moderately susceptible. Several years later, Tully and Smith[51] examined blood collected from 50 patients 24 hr postpartum by inoculating it into conventional blood culture media and onto Mycoplasma media with 20% horse serum. They recovered *M. hominis* from the cultures on Mycoplasma media only from one patient who had developed fever at this time.

Harwick et al.[52-54] reported an aggregate of six patients who became febrile either postpartum or postabortion. In one of their reports, Harwick et al.[53] outlined their blood culture techniques. In addition to inoculating blood into conventional media, they inoculated some into Mycoplasma broth containing horse serum, yeast extract, and thallium acetate. Arginine and phenol red were usually added as growth indicators for *M. hominis.* It was observed that these patients' illnesses were mild, and the authors concluded that therapy directed specifically against mycoplasmas was unnecessary and that the antibiotics of choice in postabortal patients were those active against coliforms, clostridia, and streptococci.

In two studies by the same group of investigators[55,56] in Israel, mycoplasmemias following delivery, abortion, or caesarean section were due to *M. hominis* in four patients and to *Ureaplasma urealyticum* (T strains) in two. Specific antimycoplasmal therapy was given to only one patient,[55] who responded promptly after failing to respond to various penicillins and gentamicin. One patient with negative aerobic and anaerobic cultures of a caesarean section wound infection that yielded *M. hominis* and *U. urealyticum* and whose blood cultures yielded *M. hominis* responded promptly to incision and drainage only. Penicillins and, in one case, gentamicin were used in the remaining cases without clear-cut effects.

Simberkoff and Toharsky[57] have recently reported mycoplasmemia due to *M. hominis* in five adult male patients with urinary tract obstruction, surgery, or manipulation. Two patients died, presumably of causes unrelated to their mycoplasmemia and without having received specific antimycoplasmal therapy. Two of those who recovered received specific antimycoplasmal therapy, while the third did not. The mycoplasmas in all of these cases were reported in 10 to 12 days after the cultures were inoculated and were apparently detected on blood agar subcultures that were incubated 5 to 7 days. The authors failed to reisolate mycoplasmas from blood cultures made 1 to 5 days after the initially positive ones and emphasized the transient nature of the mycoplasmemia, a feature characteristic of previously reported cases with the exception of one of three reported by Carlson et al.[46] and the one reported by Singerland and Morgan.[47]

Mycoplasmemia appears not to be a rare phenomenon in patients who become septic postpartum or following genitourinary tract manipulation or surgery. It appears that these organisms can be recovered without undue difficulty from conventional blood cultures provided that routine subcultures to Mycoplasma media are performed or that routine subcultures to blood agar are incubated 5 to 7 days and are examined carefully with a hand lens or dissecting microscope.

The clinical significance of mycoplasmemia is, however, difficult to determine. Most cases have eventually responded in the face of nonspecific antimicrobial therapy; nevertheless, two patients responded promptly to specific therapy, which consisted of incision and drainage of a mycoplasmal postoperative wound abscess in one case and administration of erythromycin in the other. Studies of large numbers of blood cultures for the presence of mycoplasmas have been unproductive.[51,53,56] There is, therefore, no clearcut indication for performing mycoplasmal blood cultures at this time, but it is important to recognize that laboratories should be able to isolate the organism in selected cases with only minor modifications of existing conventional blood culture techniques.

REFERENCES

1. Center for Disease Control, *Morbidity and Mortality Weekly Report,* (Ann. Suppl.: Summary 1975), 24, August 1976.
2. **Spink, W. W.,** *The Nature of Brucellosis,* University of Minnesota Press, Minneapolis, 1956, 464.
3. **Killough, J. H., Magill, G. B., and Said, S. I.,** Clinical and laboratory observations on Brucella melitensis in Egypt: study of 100 cases, *Ann. Intern. Med.,* 39, 222, 1953.
4. **Castaneda, M. R., Tovar, R., and Velez, R.,** Studies on brucellosis in Mexico. Comparative study of various diagnostic tests and classification of the isolated bacteria, *J. Infect. Dis.,* 70, 97, 1942.
5. **Castaneda, M. R.,** A practical method for routine blood cultures in brucellosis, *Proc. Exp. Biol. Med.,* 64, 114, 1947.
6. **McCullough, N. B.,** Laboratory tests in the diagnosis of brucellosis, *Am. J. Public Health,* 39, 866, 1949.
7. **Spink, W. W., McCullough, N. B., Hutchings, L. M., and Mingle, C. K.,** Diagnostic criteria for human brucellosis. Report no. 2 of the National Research Council, Committee on Public Health Aspects of Brucellosis, *JAMA,* 149, 805, 1952.
8. **Dalrymple-Champneys, W.,** *Brucella Infection and Undulant Fever in Man,* Oxford University Press, New York, 1960, 196.
9. **Schirger, A., Nichols, D. R., Martin, W. J., Wellman, W. E., and Weed, L. A.,** Brucellosis: experiences with 224 patients, *Ann. Intern. Med.,* 52, 827, 1960.
10. **Hamilton, P. K.,** The bone marrow in brucellosis, *Am. J. Clin. Pathol.,* 24, 580, 1954.
11. **Diesch, S. L. and Ellinghausen, H. C.,** Leptospirosis, in *Diseases Transmitted from Animals to Man,* 6th ed., Hubbert, W. T., McCulloch, W. F., and Schnurrenberger, P. R., Eds., Charles C Thomas, Springfield, Ill., 1975, 436.
12. **Ellinghausen, H. C., Jr.,** Some observations on cultural and biochemical characteristics of *Leptospira pomona, J. Infect. Dis.,* 106, 237, 1960.
13. **Galton, M. M.,** Methods in the laboratory diagnosis of leptospirosis, *Ann. N.Y. Acad. Sci.,* 98, 675, 1962.
14. **Gochenour, W. S., Jr., Yager, R. H., Wetmore, P. W., and Hightower, J. A.,** Laboratory diagnosis of leptospirosis, *Am. J. Public Health,* 43, 405, 1953.
15. **Christmas, B. W., Till, D. G., and Bragger, J. M.,** Dairy farm fever in New Zealand: isolation of L pomona and L hardjo from a local outbreak, *N. Z. Med. J.,* 79, 904, 1974.
16. **Sonnenwirth, A. C.,** The spirochetes, in *Gradwohl's Clinical Laboratory Methods and Diagnosis,* Frankel, S., Reitman, S., and Sonnenwirth, A. C., Eds., C. V. Mosby, St. Louis, 1970, 1366.
17. **Johnson, R. C. and Harris, V. G.,** Differentiation of pathogenic and saprophytic leptospires. I. Growth at low temperatures, *J. Bacteriol.,* 94, 27, 1967.
18. **Thorsteinsson, S. B., Sharp, P., Musher, D. M., and Martin, R. R.,** Leptospirosis: an underdiagnosed cause of acute febrile illness, *South. Med. J.,* 68, 217, 1975.

19. **Krick, J. A. and Remington, J. S.**, Opportunistic invasive fungal infections in patients with leukaemia and lymphoma, *Clinics in Haematology*, 5, 249, 1976.
20. **Seelig, M. S., Speth, C. P., Kozinn, P. J., Taschdjian, C. L., Toni, E. F., and Goldberg, P.**, Patterns of *Candida* endocarditis following cardiac surgery: importance of early diagnosis and therapy (an analysis of 91 cases), *Prog. Cardiovasc. Dis.*, 17, 125, 1974.
21. **Harford, C. G.**, Postoperative fungal endocarditis: fungemia, embolism, and therapy, *Arch. Intern. Med.*, 134, 116, 1974.
22. **Young, R. C., Bennett, J. E., Vogel, C. L., Carbone, P. P., and De Vita, V. T.**, Aspergillosis: the spectrum of the disease in 98 patients, *Medicine*, 49, 147, 1970.
23. **Carrizosa, J., Levison, M. E., Lawrence, T., and Kaye, D.**, Cure of *Aspergillus ustus* endocarditis on a prosthetic valve, *Arch. Intern. Med.*, 133, 486, 1974.
24. **Rubinstein, E., Noriega, E. R., Simberkoff, M. S., Holzman, R., and Rahal, J. J., Jr.**, Fungal endocarditis: analysis of 24 cases and review of the literature, *Medicine*, 54, 331, 1975.
25. **Young, R. C., Bennett, J. E., Geelhoed, G. W., and Levine, A. S.**, Fungemia with compromised host resistance: a study of 70 cases, *Ann. Intern. Med.*, 80, 605, 1974.
26. **Akbarian, M., Salfelder, K., and Schwarz, J.**, Experimental histoplasmic endocarditis, *Arch. Intern. Med.*, 114, 784, 1964.
27. **Stone, H. H., Kolb, L. D., Currie, C. A., Geheber, C. E., and Cuzzell, J. Z.**, Candida sepsis: pathogenesis and principles of treatment, *Ann. Surg.*, 179, 697, 1974.
28. **Kobza, K., Perruchoud, A., Mihatsch, J. M., and Herzog, H.**, Candidaemia and bacterial infections in patients with lung disease, *Lancet*, 2, 1084, 1976.
29. **Blazevic, D. J., Stemper, J. E., and Matsen, J. M.**, Effect of aerobic and anaerobic atmospheres on isolation of organisms from blood cultures, *J. Clin. Microbiol.*, 1, 154, 1975.
30. **Gantz, N. M., Swain, J. L., Medeiros, A. A., and O'Brien, T. F.**, Vacuum blood-culture bottles inhibiting growth of Candida and fostering growth of Bacteroides, *Lancet*, 2, 1174, 1974.
31. **Roberts, G. D., Horstmeier, C., Hall, M., and Washington, J. A., II.**, Recovery of yeast from vented blood culture bottles, *J. Clin. Microbiol.*, 2, 18, 1975.
32. **Harkness, J. L., Hall, M., Ilstrup, D. M., and Washington, J. A., II.**, Effects of atmosphere of incubation and of routine subcultures on detection of bacteremia in vacuum blood culture bottles, *J. Clin. Microbiol.*, 2, 296, 1975.
33. **Braunstein, H. and Tomasulo, M.**, A quantitative study of the growth of *Candida albicans* in vented and unvented blood-culture bottles, *Am. J. Clin. Pathol.*, 66, 87, 1976.
34. **Roberts, G. D. and Washington, J. A., II.**, Detection of fungi in blood cultures, *J. Clin. Microbiol.*, 1, 309, 1975.
35. **Roberts, G. D., Horstmeier, C. D., and Ilstrup, D. M.**, Evaluation of a hypertonic sucrose medium for the detection of fungi in blood cultures, *J. Clin. Microbiol.*, 4, 110, 1976.
36. **Parsons, R. J. and Zarafonetis, C. J. D.**, Histoplasmosis in man: report of seven cases and a review of seventy-one cases, *Arch. Intern. Med.*, 75, 1, 1945.
37. **Portnoy, J., Wolf, P. L., Webb, M., and Remington, J. S.**, Candida blastospores and pseudohyphae in blood smears, *N. Engl. J. Med.*, 285, 1010, 1971.
38. **Anderson, A. O. and Yardley, J. H.**, Demonstration of Candida in blood smears, *N. Engl. J. Med.*, 286, 108, 1972.
39. **Silverman, E. M., Norman, L. E., Goldman, R. T., and Simmons, J.**, Diagnosis of systemic candidiasis in smears of venous blood stained with Wright's stain, *Am. J. Clin. Pathol.*, 60, 473, 1973.
40. **Humphrey, A. A.**, Use of the buffy layer in the rapid diagnosis of septicemia, *Am. J. Clin. Pathol.*, 14, 358, 1944.
41. **Komorowski, R. A. and Farmer, S. G.**, Rapid detection of candidemia, *Am. J. Clin. Pathol.*, 59, 56, 1973.
42. **Sullivan, N. M., Sutter, V. L., and Finegold, S. M.**, Practical aerobic membrane filtration blood culture technique: clinical blood culture trial, *J. Clin. Microbiol.*, 1, 37, 1975.
43. **Sullivan, N. M., Sutter, V. L., and Finegold, S. M.**, Practical aerobic membrane filtration blood culture technique: development of procedure, *J. Clin. Microbiol.*, 1, 30, 1975.
44. **Dorn, G. L., Haynes, J. R., and Burson, G. G.**, Blood culture technique based on centrifugation: developmental phase, *J. Clin. Microbiol.*, 3, 251, 1976.
45. **Dorn, G. L., Burson, G. G., and Haynes, J. R.**, Blood culture technique based on centrifugation: clinical evaluation, *J. Clin. Microbiol.*, 3, 258, 1976.
46. **Carlson, H. J., Spector. S., and Douglas, H. G.**, Possible role of pleuropneumonia-like organisms in etiology of disease in childhood, *Am. J. Dis. Child.*, 81, 193, 1951.
47. **Singerland, D. W. and Morgan, H. R.**, Sustained bactermia with pleuropneumonia-like organisms in a postpartum patient, *JAMA*, 150, 1309, 1952.
48. **Stokes, E. J.**, Human infection with pleuropneumonia-like organisms, *Lancet*, 1, 276, 1955.
49. **Stokes, E. J.**, PPLO in genital infections, *Br. Med. J.*, 1, 510, 1959.
50. **Tully, J. G., Brown, M. S., Sheagren, J. N., Young, V. M., and Wolff, S. M.**, Septicemia due to *Mycoplasma hominis* type 1, *N. Engl. J. Med.*, 273, 648, 1965.
51. **Tully, J. G. and Smith, L. G.**, Postpartum septicemia with *Mycoplasma hominis, JAMA*, 204, 827, 1968.
52. **Harwick, H. J., Iuppa, J. B., Purcell, R. H., and Fekety, F. R., Jr.**, *Mycoplasma hominis* septicemia associated with abortion, *Am. J. Obstet. Gynecol.*, 99, 725, 1967.

53. **Harwick, H. J., Purcell, R. H., Iuppa, J. B., and Fekety, F. R., Jr.,** *Mycoplasma hominis* and abortion, *J. Infect. Dis.*, 121, 260, 1970.
54. **Harwick, H. J., Purcell, R. H., Iuppa, J. B., and Fekety, F. R., Jr.,** *Mycoplasma hominis* and postpartum febrile complications, *Obstet. Gynecol.*, 37, 765, 1971.
55. **Caspi, E., Herczeg, E., Solomon, F., and Sompolinsky, D.,** Amnionitis and T strain mycoplasmemia, *Am. J. Obstet. Gynecol.*, 111, 1102, 1971.
56. **Solomon, F., Caspi, E., Bukovsky, I., and Sompolinsky, D.,** Infections associated with genital Mycoplasma, *Am. J. Obstet. Gynecol.*, 116, 785, 1973.
57. **Simberkoff, M. S. and Toharsky, B.,** Mycoplasmemia in adult male patients, *JAMA*, 236, 2522, 1976.

Chapter 4

ANTIMICROBIAL SUSCEPTIBILITY TESTS

John A. Washington II

The rapid performance of in vitro susceptibility tests with blood-culture isolates provides the clinician with essential information for the appropriate selection of an antibiotic and the dosage required for the therapy of septicemia. Those organisms most frequently involved in septicemia are also those which most frequently display unpredictable susceptibility to various antibiotics and do, therefore, require susceptibility testing as expeditiously as possible. These organisms include the staphylococci, the Enterobacteriaceae, and *Pseudomonas.* Furthermore, groups of bacteria, such as *Haemophilus influenzae,* previously considered to be uniformly susceptible to certain antibiotics, no longer are uniformly susceptible and require prompt testing on an individual basis to determine this fact.

Although there is little question about testing *Staphylococcus aureus,* the Enterobacteriaceae, and *Pseudomonas* when isolated from the blood, questions often arise about the necessity of testing *S. epidermidis, Bacillus,* diphtheroids, and viridans streptococci, all of which when present only in single sets of cultures ordinarily represent contaminants or are of little or no clinical significance. Policies on this matter will inevitably differ from hospital to hospital, and there is no universal rule of thumb to follow. Procedures at the Mayo Clinic dictate that unless such bacteria are present in multiple sets of cultures, routine susceptibility tests are not performed. However, at least two separate sets of blood cultures are routinely collected from each patient within a 24-hr period so that a determination that multiple sets (at least two bottles per set) of cultures are positive can be easily made. In such cases, the patient's physician is called to determine the possible clinical significance of these cultures and, if they are deemed to be significant, which antibiotics should be tested. An active infectious diseases service is in operation at the Clinic also and is notified about all positive blood cultures. Each case, therefore, is carefully considered on its own merits.

Selection of Antimicrobial Agents for Testing

Those agents that should be selected for testing against blood-culture isolates are listed in Table 1. Although a single agent from each class of agents usually suffices for routine purposes, the variety of

TABLE 1

Selection of Antimicrobial Agents to be Tested Routinely Against Bacterial Blood-Culture Isolates

Antibiotic	Staphylococci	Streptococci	Enterobacteriaceae	*Pseudomonas*
Amikacin			X	X
Ampicillin		X	X	
Carbenicillin				X
Cephalothin	X	X	X	
Chloramphenicol[a]	X		X	
Clindamycin	X			
Erythromycin	X			
Gentamicin	X[a]		X	X
Kanamycin	X[a]		X	X
Methicillin[b]	X			
Penicillin G	X	X		
Tetracycline			X	
Tobramycin			X	X
Vancomycin[a]	X			

[a]To be tested as secondary agent only.
[b]Acceptable substitutes are oxacillin and nafcillin.

penicillins with differing spectra of activity and the advent of new aminoglycosides providing varying substrates for aminoglycoside inactivating enzymes make the testing of more than one representative of each of these classes of antibiotics necessary. The proliferation of newer cephalosporins, such as cefoxitin and cefamandole, further compounds the laboratory's problems with in vitro susceptibility testing. Again, practices will vary from hospital to hospital and will, to a great extent, reflect local problems with in vitro susceptibility of organisms to particular antibiotics. Currently, Mayo Clinic procedure is to test amikacin routinely against Gram-negative bacilli, along with gentamicin, but to report its minimal inhibitory concentrations (MIC) only in those few instances where gentamicin MIC values exceed 2 μg/ml.

Tests

There are a variety of tests which are germane to bacteremia. They will be considered here in principle; however, the reader is referred elsewhere for the technical details of their performance.[1]

Dilution

The lowest or minimal concentration (μg/ml) of antimicrobial agent required to inhibit the growth of an organism can be determined by inoculating a standardized quantity of the organism into tubes containing broth or plates containing agar that also contain a graded series of concentrations of the agent.[2] After overnight incubation, the MIC is the lowest concentration of antimicrobial agent which completely inhibits the organism's growth. Although the reference method for determining the MIC specifies a twofold or serial dilution scheme for the antimicrobial agent in broth or agar, it is practical on a routine basis to test fewer concentrations in an expanded dilution scheme in agar.[1] Alternatively, there are mechanical devices available for preparing microdilutions of the antimicrobial agents in broth. Both systems have their advantages and disadvantages but are distinctly more efficient than manually performing macrodilutions in broth on an individual basis.

In those few instances in which determining minimal bactericidal concentrations are indicated, broth (macro or micro) dilution tests are advantageous.

The objective of performing dilution tests is to obtain quantitative results, an approximation of which can be made by extrapolating zone diameters obtained from diffusion tests (see below) to an MIC value on a regression graph, providing that one has access to the zone diameter and the regression graph for that particular agent. Additionally, the disc diffusion test should be performed by the standardized method. In general, however, the specific indications for performing dilution tests are as follows:

1. Tests on slow-growing, fastidious, or anaerobic bacteria for which standardized disc diffusion methods do not exist
2. Confirmation of susceptibility to the polymyxins which diffuse poorly in agar and are not reliably tested by the disc diffusion method
3. Confirmation of resistance to aminoglycosides and, more specifically, to gentamicin and amikacin for which the disc diffusion method may provide an erroneously resistant result
4. Tests on bacteria which are found to be intermediately susceptible to potentially toxic antibiotics but for which there are no acceptable alternatives
5. Tests on certain urinary isolates which are resistant by disc diffusion interpretative criteria, but are preferentially treated with the higher levels of penicillins and cephalosporins attainable in the urine
6. Determination of bactericidal activity
7. As an alternative to the disc diffusion method on a routine basis.

Knowledge of the degree of susceptibility of an organism to a particular antibiotic may be very helpful in the selection of the appropriate agent for therapy. An isolate of *P. aeruginosa* which is inhibited by 2 μg/ml of gentamicin would be amenable to therapy with gentamicin; however, if its MIC value were 4 or 8 μg/ml, tobramycin or amikacin might be preferable, providing, of course, that their MIC values were within the susceptible ranges.

The provision of MIC values does impose a considerable educational responsibility on the laboratory and on a clinician interested in infectious diseases, in that the proper utilization of this value depends to a great extent on one's familiarity with the pharmacokinetics of various antimicrobial agents. Obviously, many factors influence one's choice of an antimicrobial agent and its dosage, but the MIC provides an added bit of sophistication to the process which most physicians have not been trained to utilize. While susceptibility is generally considered to occur when the MIC is one half to one quarter of the safely achievable peak serum level of the antibiotic, one must know what this peak level is and what its variability is, depending on the antibiotic's route of administration.

Diffusion

For most purposes, the determination of in vitro susceptibility can be performed by the standardized disc diffusion method, originally described by Bauer et al.[3] This procedure has subsequently been specified as the standard method by the U.S. Food and Drug Administration and more recently described in greater detail in an approved standard by the National Committee for Clinical Laboratory Standards. As in any antimicrobial susceptibility test, standardization of the inoculum, the medium, and other aspects of the procedure is essential to the production of reliable and reproducible results.

Until recently in both the dilution and diffusion methods, it has been considered necessary to use colonies obtained in subcultures to prepare the inoculum. Wegner et al.[4] have described a direct method of susceptibility testing wherein broth suspensions of positive blood cultures containing either staphylococci or Gram-negative bacilli were incubated at 37°C for 4 to 6 hr, following which the turbidity of the suspensions was adjusted with 0.9% NaCl to match that of half of a McFarland No. 1 $BaSO_4$ standard before inoculating them onto the surface of Mueller-Hinton agar plates and applying discs in the standard manner. With 74 positive cultures containing single organisms, there were only five instances (0.7%) in which a change in the interpretation of the results was required.

In a similar comparison of direct and standardized susceptibility testing of positive blood cultures using an agar dilution technique, Johnson and Washington demonstrated an overall agreement of 87.9% in 1536 antibiotic comparisons.[5] Agreement between the two methods was 98.3% if minor discrepancies were disregarded. In this study, an attempt was also made to adjust the turbidity of the broth removed from the positive culture to match that of the one half of the McFarland No. 1 standard by adding a sample of the broth to 2 ml of Mueller-Hinton broth and incubating the mixture at 35°C for 2 to 4 hr.

Direct susceptibility tests, therefore, are both feasible and accurate for positive blood cultures, providing that the cultures contain rapidly growing aerobic or facultatively anaerobic bacteria in pure culture, the inoculum is standardized, and that the remainder of the procedure is also standardized.

Bactericidal

Bactericidal tests are frequently performed with blood-culture isolates and especially with those from patients with infective endocarditis. Three types of tests are involved: the determination of the minimal bactericidal concentration (MBC) of a single antibiotic, the determination of the antibacterial activity of antibiotics in combination, and the determination of the serum bactericidal titer (SBT or Schlichter test).

The principle of the test for the MBC is to determine the lowest concentration of antibiotic that will kill 99.9 to 100% of a standardized inoculum of the patient's organism. In performing the test, therefore, it is necessary to quantitate the original inoculum's size to determine the percentage that is destroyed within the temporal limits and conditions of the test. There are two general approaches to the test. In one, a standard overnight incubation, identical to that used for the determination of the MIC in broth, is performed with quantitative subcultures of those tubes containing broth without visible evidence of growth. In the Mayo Clinic laboratory, subcultures of rapidly growing organisms are made in pour plates, while those of more slowly growing or fastidious organisms are made in thioglycollate medium.[1] The usual presence of a small quantity of agar in thioglycollate media enables one to enumerate small numbers of suspended colonies, providing that the tubes are well mixed prior to their incubation.

Most of the determinations of MBC for blood-culture isolates in the Clinic laboratory are with streptococci from patients with infective endocarditis. In nearly all of these instances, subcultures to thioglycollate medium are necessary for purposes of quantitating survivorship.

The other general approach to bactericidal testing is the timed-killing curve whereby subcultures of tubes of broth containing one or more fixed concentrations of antibiotic are made at several intervals during incubation to determine

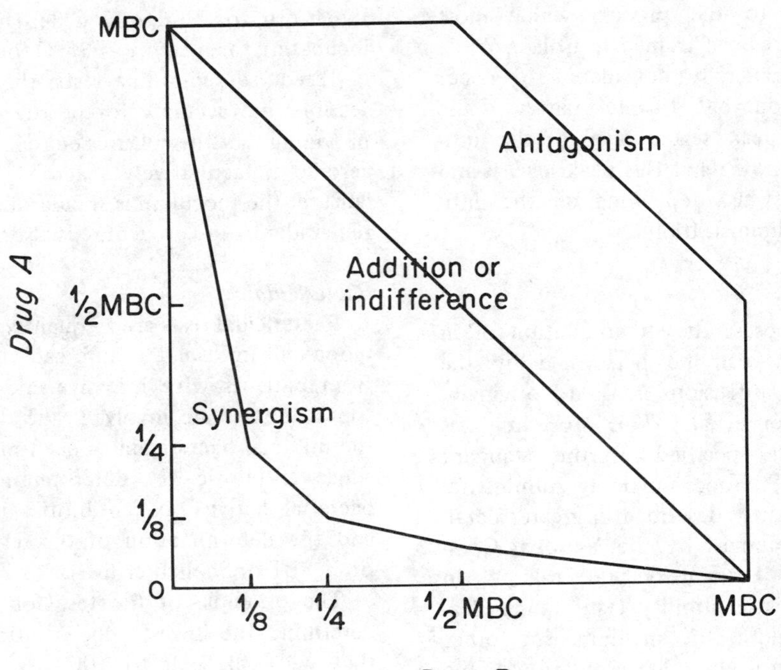

FIGURE 1. Isobols representing three types of interactions between drugs A and B.

the rate of killing. This method will be dealt with in more detail below, as studies of the bactericidal activity of combinations of antimicrobial agents are discussed.

The method of performing the MBC has not been standardized. As with any antimicrobial susceptibility test, several variables such as the original inoculum size and the medium used initially and for subcultures do influence the outcome of the test. Marcoux and Washington[6] found the bactericidal activity of methicillin against *S. aureus* to vary considerably according to the medium employed for the subcultures. No bactericidal activity was noted if subcultures were made in thioglycollate medium from either soybean-casein digest broth or brain heart infusion broth. In contrast, significant bactericidal activity was noted if the subcultures were made on blood agar plates or from Mueller–Hinton broth into thioglycollate medium or onto blood agar plates. These medium-related problems are, of course, to be distinguished from the so-called penicillin-tolerant strains of *S. aureus* described recently by Sabath et al.[7] These strains, isolated from the blood or bone cultures of several patients responding poorly to antistaphylococcal therapy, have low MIC values but high MBC values irrespective of the medium employed. These strains appear to be penicillin-tolerant by virtue of decreased autolytic enzyme activity.

Bactericidal tests of combinations of antimicrobial agents are being performed with increasing frequency against resistant blood-culture isolates from patients with infective endocarditis and with bacteremias of other etiologies. There are two basic approaches to the performance of these tests. In the two-dimensional checkerboard technique, originally described by Loewe[8] and subsequently modified by Lacey[9] and Sabath,[10] serial, twofold dilutions of two antimicrobial agents (alone and in combination) are inoculated with a standardized inoculum of the test organism.[1] After overnight incubation, those tubes containing broth with no visible turbidity are subcultured as for the determination of the MBC. The results of this test are expressed according to isobologram criteria as demonstrating antagonistic, additive, or synergistic effects (Figure 1).

The other approach to testing the in vitro effects of combinations of antibiotics is the timed-killing curve wherein fixed concentrations of each drug alone and in combination with broth

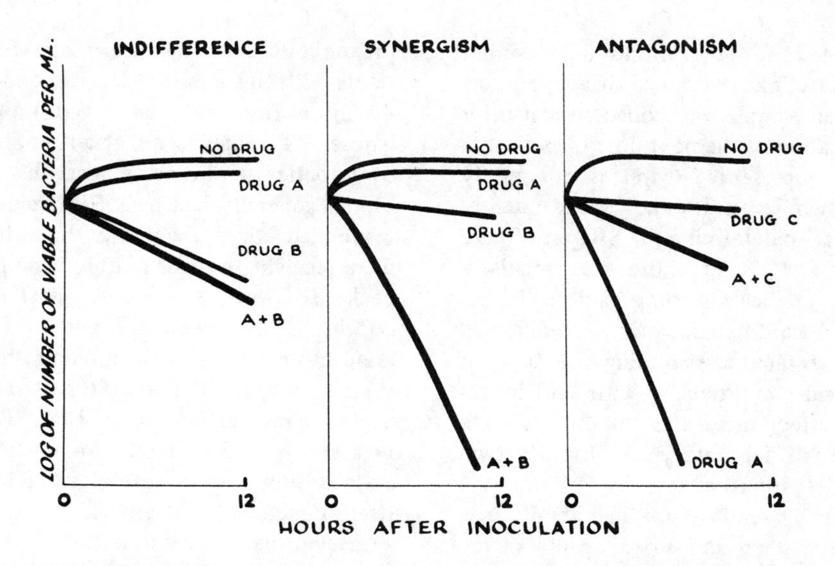

FIGURE 2. Schematic representation of bactericidal action in vitro showing the possible types of results seen when one or two drugs act on a homogeneous population of bacteria under conditions permitting growth. (From Jawetz, E., *Antimicrob. Agents Chemother.*, 1967, 203, 1968. With permission.)

are inoculated with standardized suspensions of microorganisms and are subsequently subcultured to determine their bactericidal activity at varying time intervals. Indifferent, synergistic, or antagonistic effects of such drug combinations are then defined according to the criteria set forth by Jawetz[11] (Figure 2). In this instance, synergism implies both the increased bactericidal activity of drugs tested in combination relative to their activity when tested singly and the ability of two drugs in combination to increase the rate of bactericidal activity relative to the rate with either drug singly.

In the two-dimensional checkerboard technique, subcultures of broth-antibiotic mixtures containing no visible evidence of growth are generally obtained one time, usually after 18 to 24 hr of incubation. In the timed-killing curve, multiple subcultures are made, usually beginning after 4 to 6 hr of incubation. It is essential in the timed-killing curve to perform quantitative subcultures to ascertain both the extent and the rate of bactericidal activity of the drugs being tested.

Although there are strong advocates for each of these two approaches for testing the effects of combinations of antibiotics, there is usually (as was pointed out by Libke et al.[12] and Weinstein et al.[13]) a complete correlation between the two in terms of determining the presence of synergism, addition, or indifference. Both procedures are, unfortunately, equally tedious and time-consuming to perform.

Numerous reports have confirmed the clinical value of the combined use of penicillins and other cell-wall active antibiotics with aminoglycosides in the treatment of enterococcal group D streptococcal endocarditis and of certain Gram-negative infections; however, more data are needed to document the correlation between combination studies in vitro and the clinical response in vivo. The clinical significance, for example, of the lack of synergy in vitro in many instances between penicillins and streptomycin against enterococcal group D streptococci[12,14,15] remains unclear; however, Klastersky et al.[16] found that in 148 severe episodes of Gram-negative infections in patients with disseminated cancer there was a significantly better response to the use of antibiotic combinations that were synergistic in vitro against the etiologic agent than to combinations that were not synergistic in vitro. Similar data were cited by Weinstein et al.[13] The clinical value of more recent reports demonstrating synergism in vitro between semisynthetic penicillins and gentamicin against *S. aureus* remains to be established.

The determination of the serum inhibitory titer (SIT) was described by Schlichter and coworkers[17] in 1949. They felt that a penicillin serum level producing complete inhibition in a 1:4 dilution of serum represented the optimal dosage to be maintained for the duration of either bacteriological or clinical evidence of active infection. Their serum sample was collected just prior to the administration of the next dose of penicillin and, therefore, represented what is commonly termed the trough level. Jawetz[18] subsequently modified the test to determine the SBT and found that an SBT of 1:10 or more was regularly associated with eradication of organisms in patients with endocarditis. Jawetz[18] emphasized that a variety of factors, including the type of organism, cultural conditions, volume and inoculum size, may affect the test results and recommended that each laboratory develop its own confidence limits for the test by relating the methodology to the end results in patients in a particular hospital. Pien and Vosti,[19] in a survey of 37 clinical laboratories, did indeed find substantial variability in the manner of performance of this test. Pien et al.[20] subsequently reported that the use of serum rather than broth as the diluent in the serum bactericidal test resulted in a significant decrease in the SBT among patients receiving highly protein-bound semisynthetic penicillins and unaccountably recommended that serum be used as the standard diluent for the test, despite the fact that Jawetz's[18] rather widely recommended minimum acceptable titer of 1:10 was based on the dilution of serum in broth.

There have been remarkably few studies since those reported by Jawetz[18] which have addressed themselves to the clinical significance of the SBT. Klastersky et al.[21] determined the peak and trough SIT and SBT values of the serum of 317 patients with cancer and with bacteriologically proven infectious diseases. They found that patients with SIT values ⩾1:8 responded satisfactorily more frequently ($p = 0.01$) than did those whose SIT values never attained this level. Moreover, patients who were receiving multiple antibiotics generally had peak SBT values of 1:16 and responded more favorably than did those who were receiving a single antibiotic and generally had peak SBT values of 1:4. Klastersky et al.[21] concluded that peak SIT and SBT values correlated more clearly with outcome than did trough values and that the inhibitory titers were more useful than the bactericidal titers. Bryan et al.,[22] in a study of the SBT in Gram-negative bacillary endocarditis and of several variables of the test itself, concluded that a peak titer of less than 1:8 suggested inadequacy of antimicrobial therapy and that if toxicity occurred with titers consistently exceeding 1:16, the SBT served as a useful guide for evaluating therapy while reducing dosage, modifying the route of administration of antimicrobial agents, or altering the number or types of agents used.

In conclusion, for most purposes, determination of the susceptibility of a blood-culture isolate by the disc diffusion or by a dilution procedure suffices. In certain instances, however, the determination of the MBC can be helpful in the proper selection of antimicrobial agents for therapy, and in a few cases with resistant microorganisms, studies of the combined effects of agents may be quite useful. The serum bactericidal titer appears to be of value in selected cases and especially in those with infective endocarditis in monitoring the adequacy of therapy.

REFERENCES

1. **Washington, J. A., II, Ed.,** *Laboratory Procedures in Clinical Microbiology,* Little, Brown, Boston, 1974.
2. **Washington, J. A., II and Barry, A. L.,** Dilution test procedures, in *Manual of Clinical Microbiology,* 2nd ed., Lennette, E. H., Spaulding, E. H., Truant, J. P., Eds., American Society for Microbiology, Washington, D.C., 1974, 410.
3. **Bauer, A. W., Kirby, W. M. M., Sherris, J. C., and Turck, M.,** Antibiotic susceptibility testing by a standardized single disc method, *Am. J. Clin. Pathol.,* 45, 493, 1966.
4. **Wegner, D. L., Mathis, C. R., and Neblett, T. R.,** Direct method to determine the antibiotic susceptibility of rapidly growing blood pathogens, *Antimicrob. Agents Chemother.,* 9, 861, 1976.
5. **Johnson, J. E. and Washington, J. A., II,** Comparison of direct and standardized antimicrobial susceptibility testing of positive blood cultures, *Antimicrob. Agents Chemother.,* 10, 211, 1976.
6. **Marcoux, J. A. and Washington, J. A., II,** Pitfalls in identification of methicillin-resistant *Staphylococcus aureus, Appl. Microbiol.,* 18, 699, 1969.
7. **Sabath, L. D., Wheeler, N., Laverdiere, M., and Blazevic, D.,** A new type of penicillin resistance of *Staphylococcus aureus, Lancet,* 1, 443, 1977.

8. **Loewe, S.**, The problem of synergism and antagonism of combined drugs, *Arzneim. Forsch.*, 3, 285, 1953.
9. **Lacey, B. W.**, Mechanisms of chemotherapeutic synergy, in *The Strategy of Chemotherapy*, Cowan, S. T. and Rowatt, E., Eds., Cambridge University Press, New York, 1958, 247.
10. **Sabath, L. D.**, Synergy of antibacterial substances by apparently known mechanisms, *Antimicrob. Agents Chemother.*, 1967, 210, 1968.
11. **Jawetz, E.**, Combined antibiotic action: some definitions and correlations between laboratory and clinical results, *Antimicrob. Agents Chemother.*, 1967, 203, 1968.
12. **Libke, R. D., Regamey, C., Clarke, J. T., and Kirby, W. M. M.**, Synergism of carbenicillin and gentamicin against enterococci, *Antimicrob. Agents Chemother.*, 4, 564, 1973.
13. **Weinstein, R. J., Young, L. S., and Hewitt, W. L.**, Comparisons of methods for assessing in vitro antibiotic synergism against *Pseudomonas* and *Serratia*, *J. Lab. Clin. Med.*, 86, 853, 1975.
14. **Wilkowske, C. J., Facklam, R. R., Washington, J. A., II, and Geraci, J. E.**, Antibiotic synergism: enhanced susceptibility of group D streptococci to certain antibiotic combinations, *Antimicrob. Agents Chemother.*, 1970, 195, 1971.
15. **Moellering, R. C., Jr., Wennersten, C., and Weinberg, A. N.**, Synergism of penicillin and gentamicin against enterococci, *J. Infect. Dis.*, 124, S207, 1971.
16. **Klastersky, J., Cappel, R., and Daneau, D.**, Clinical significance of in vitro synergism between antibiotics in gram-negative infections, *Antimicrob. Agents Chemother.*, 2, 470, 1972.
17. **Schlichter, J. G., MacLean, H., and Milzer, A.**, Effective penicillin therapy in subacute bacterial endocarditis and other chronic infections, *Am. J. Med. Sci.*, 217, 600, 1949.
18. **Jawetz, E.**, Assay of antibacterial activity in serum, *Am. J. Dis. Child.*, 103, 81, 1962.
19. **Pien, F. D. and Vosti, K. L.**, Variation in performance of the serum bactericidal test, *Antimicrob. Agents Chemother.*, 6, 330, 1974.
20. **Pien, F. D., Williams, R. D., and Vosti, K. L.**, Comparison of broth and human serum as the diluent in the serum bactericidal test, *Antimicrob. Agents Chemother.*, 7, 113, 1975.
21. **Klastersky, J., Daneau, D., Swings, G., and Weerts, D.**, Antibacterial activity in serum and urine as a therapeutic guide in bacterial infections, *J. Infect. Dis.*, 129, 187, 1974.
22. **Bryan, C. S., Marney, S. R., Jr., Alford, R. H., and Bryant, R. E.**, Gram-negative bacillary endocarditis: interpretation of the serum bactericidal test, *Am. J. Med.*, 58, 209, 1975.

Chapter 5
NEW OR EXPERIMENTAL APPROACHES TO DETECTION OF BACTEREMIA

John P. Anhalt

INTRODUCTION

The current emphasis on development of new methods in microbiology is due in part to the need for more timely results to make optimal use of modern antimicrobial agents, and in part to increased demand for improving laboratory function and lowering laboratory costs for patient care. Several recent reviews have considered automation in microbiology as a general theme.[1-3] The purpose of this chapter is to review the application of newer methods specifically to detection of bacteremia. Techniques that use centrifugation or filtration to recover bacteria from blood are described separately. These techniques represent newer methods to process a blood specimen before culture. In the second section, newer methods to detect bacterial growth are described. These methods have been used in conjunction with more or less conventional blood culture systems, as well as with blood culture systems that involve filtration techniques. Finally, methods based on detection of bacterial products are described. Since bacterial products produced at the site of an infection may be released into the circulatory system and detected in the blood, the procedures reviewed in this section are not strictly for detection of bacteremia. Rather than cover each of these procedures in depth, a survey is provided emphasizing the relationship of bacteremia to the detection of bacterial products in blood.

FILTRATION/CENTRIFUGATION

Less than optimal conditions for the growth of bacteria may result when blood from a patient with suspected bacteremia is added directly to a culture medium. Various inhibitory factors present in blood, including antimicrobial drugs, can prevent or delay bacterial growth and detection. Additional inhibitory substances are released during coagulation of whole blood.[4] Adequate dilution of the blood and the addition of anticoagulants and antimicrobial antagonists have been shown to reduce these effects.[5,6] These measures, however, are not entirely adequate to eliminate the inhibitory effects of all antimicrobial agents that may be present in a blood specimen obtained for culture. Sullivan et al.[7] and Kagan et al.[8] have shown that during antibiotic therapy, conventional broth cultures may show no growth, while cultures processed by the techniques to be described below remain positive.

Conceptually, inhibition by the various factors in blood could be avoided by concentration or isolation of the bacteria prior to culture. The earliest attempts to apply this concept can be traced to Pickett and Nelson[9] in 1951. These investigators used centrifugation of laked blood to concentrate *Brucella* for culture. Their observation that *Brucella* could be recovered from most blood specimens, including blood from normal individuals, was not confirmed by subsequent investigators. Shortly after membrane filters became commercially available, Braun and Kelsh,[10] in 1954, developed a filtration technique to recover *Brucella* from blood of experimentally infected rabbits. Subsequent studies have used filtration techniques, and only recently has centrifugation been reinvestigated as a method to concentrate bacteria in blood.[11,12]

The principle of filtration techniques is that bacteria can be retained by a filter which allows the inhibitory substances in blood to pass through. The retained bacteria may be washed to remove residual inhibitory substances and can then be cultured on appropriate media. The filter may be placed in a broth that allows growth of the bacteria, but more commonly the filter is placed on an agar medium with the upstream side away from the agar surface. Nutrients from the medium diffuse through the filter and allow the captured bacteria to grow as discrete colonies on the filter surface. Different media or conditions of incubation may be used with a single filter by dividing it accordingly. To filter blood to recover bacteria, the formed elements, principally erythrocytes, must be removed or lysed to prevent filter clogging. Various techniques have been used to facilitate filtration, and the major developments have been in this area.

The early studies used various methods to lyse

erythrocytes in order to obtain a filtrable mixture from which bacteria could be recovered. These procedures were cumbersome and allowed recovery of bacteria from only 1 to 2 ml of blood. Nevertheless, the filtration technique had certain advantages over the broth techniques that were used for comparison. Tidwell and Gee[13] found that a filtration technique was more sensitive and allowed more rapid detection of experimental bacteremia in rabbits. Vacek and Svejcar[14] used a filtration technique in a clinical study and found advantages for detection of staphylococcal sepsis and bacteremia in patients receiving antibiotic therapy. Dodin et al.[15] concluded that in addition to more rapid detection, the discrete colonies formed on the filter allowed quantitation and more rapid identification of Gram-negative bacteremia, since differential or selective media could be used for primary isolation. Randriambololona and Dodin[16] later showed that although conventional blood cultures in typhoid fever rapidly become negative with chloramphenicol therapy, bacteria could be isolated for several additional days by using a filtration technique.

More recent studies have used improved techniques for blood culture and have developed procedures to recover bacteria from up to 25 ml of blood.[17] This latter factor is of particular importance in the detection of low-level bacteremia, which may be missed simply due to the chance distribution of bacteria.[7,12,17,18] These recent experiments are summarized in Table 1, which is arranged chronologically. Each procedure is classified by the method that was used to process blood for filtration or centrifugation. In sedimentation filtration, bacteria-rich plasma was prepared by allowing the erythrocyte mass to sediment in a dextran[18,19] or glucose[17] medium or by centrifugation of anticoagulated blood at low relative forces (310 × *g*).[20] The bacteria were then collected by filtration and cultured. In some experiments, leukocytes were separated from diluted plasma by centrifugation[18,19] or by selective filtration with filters that allowed bacteria, but not leukocytes, to pass.[17] Separate culture of the leukocyte fraction should have allowed differentiation of extracellular bacteremia from leukocyte-associated infection; however, antibiotic therapy interfered with the ability to make this distinction.[18,19] For ordinary clinical purposes, separate culture of the leukocyte fraction was considered unnecessary.[18] Culture of the erythrocyte mass did not significantly increase the number of positive cultures and was also unnecessary.[17-19]

In lysis filtration, anticoagulated whole blood was subjected to lysis by various chemical and enzymatic agents. Bacteria were then recovered from the lysed specimen by filtration. The filter was placed either on solid media[7,21-24] or in broth[8,25] for culture. The solutions originally used for lysis contained relatively high concentrations of Triton X-100® (Rohm and Haas, Philadelphia, Pennsylvania) and Na_2CO_3.[21,26] Farmer and Komorowski[21] showed that a solution containing 0.025% of Triton X-100 and 0.04% of Na_2CO_3 was toxic for a wide variety of bacteria (Table 1). Lower concentrations of these reagents were significantly less toxic and probably did not affect recovery of Gram-negative or fastidious bacteria.[7,24,25,27] Schrot et al.[27] and Sullivan et al.[24] suggested that bacterial toxicity of solutions containing Triton X-100 may be more related to pH than to the ingredients themselves. Zierdt et al.[25] used 0.04% Triton X-100 and 0.6% Rhozyme® with a sodium carbonate buffer to give a final pH of 7.8 and observed acceptable levels of toxicity for several species of fastidious bacteria. This experiment indicated that pH was a significant factor, since the concentration of Triton X-100 was higher than that used by Farmer and Komorowski.[21] Filter clogging was a problem when blood with an elevated leukocyte count was lysed.[7,25] This problem was partially solved by the addition of an enzyme preparation to the lysing solution. Varidase® (Lederle Laboratories, Pearl River, New York)[7] and Rhozyme 41® (Rohm and Haas, Philadelphia, Pennsylvania)[25,27] have been used (Table 1), but a comparison of the two preparations for this purpose has not been reported. (Additional factors that have contributed to poor filtration are described in Table 1.)

In lysis centrifugation, the erythrocytes in anticoagulated whole blood were lysed, and bacteria and leukocytes were concentrated on or into a stabilizing density layer prepared from a sucrose-gelatin mixture. This layer was then removed and plated onto solid agar.[11,12]

Evident in Table 1 is the wide variety of techniques that have been used to concentrate or isolate bacteria from blood. Since various comparison methods were also used in these studies, accurate assessment of the relative merit of the different techniques for detection of bacteremia is

TABLE 1

Comparison of Recent Methods for Concentration of Bacteria Present in Blood

Method	Filter type (pore size, diameter, supplier)[a]	Solution used for sedimentation or lysis[b]	Effective blood volume treated	Results (positive cultures): Broth comparison[c,d]	Centrifugation/filtration[c]	Total[e]	Remarks	Ref.
Sedimentation filtration	0.45 μm; 37 mm; NS	3% dextran; 0.6% NaCl; 1% sodium citrate; 20 ml	3–10 ml	21(2)	23(4)	25(4)	Filter rapidly clogged and resulted in antibiotic carry-over	19
Sedimentation filtration	0.45 μm and 0.8 μm; 37, 47, and 90 mm; NS	NS	10 ml	NS	NS	NS	More rapid and complete filtration by using larger filter and/or larger pore size; no decrease in bacterial recovery with a pore size of 0.8 μm	22
Sedimentation filtration	0.45 μm and 3.0 μm; NS; M	5% glucose in a broth medium; pH 6.4 to 6.6; 250 ml	12–15 ml	24(1)	26(3)	27(2)	Did not wash filtered materials; filters used in series; filtration rapid and complete in less than 2 minutes	17
Lysis filtration	0.45 μm; 47 mm; M	0.05% Triton X-100®, 0.08% Na_2CO_3	4.2 ml	68 (NS)	NS (25)	93 (2)	Volume of lysing solution not reported; blood collected in a Vacutainer® tube containing SPS as anticoagulant; comparison method presumably used a single bottle containing thioglycollate broth	26
Sedimentation filtration	0.45 μm; 37 mm; NS	3% dextran, 0.6% NaCl, 1% sodium citrate; 20 ml	3 ml	55 (NS)	46 (NS)	66 (6)	Results from leukocyte culture not included in filtration results; filter rapidly clogged; probably antibiotic carryover; technique cumbersome and required 2 hr per culture for processing	18
Sedimentation filtration	0.45 μm; NS; M	0.35% SPS; 1.7 ml	8.3 ml	9	9	9 (2)	Erythrocytes separated by centrifugation at 310 × *g* in a Vacutainer tube containing SPS as anticoagulant	20

[a]Abbreviations used: NS, not specified; G, Gelman, Ann Arbor, Mich.; M., Millipore, Bedford, Mass.; NG, Nalge, Rochester, N.Y.
[b]Trademarks: Vacutainer®, Becton-Dickinson and Co., Rutherford, N.J.; Triton X-100®, Rhozyme 41®, Rohm and Haas, Philadelphia, Pa.; Varidase®, Lederle Laboratories, Pearl River, N.Y.; Solryth®, Medical Research, Dallas, Tex.
[c]Number in parentheses represents the number of positive cultures by each method exclusively.
[d]Broth comparison includes data from usually two, 100- or 50-ml broth cultures in which a blood dilution of 1:10 or 1:20 was used. Exceptions are noted.
[e]In many instances, more than one comparison method was used, such as pour plates or osmotically stabilized broth. The total represents the number of positive cultures by one or more of the methods used. The number in parentheses represents the number of methods from which data are included.

TABLE 1 (continued)

Comparison of Recent Methods for Concentration of Bacteria Present in Blood

Method	Filter type (pore size; diameter; supplier)[a]	Solution used for sedimentation or lysis[b]	Effective blood volume treated	Results (positive cultures): Broth comparison[c,d]	Results (positive cultures): Centrifugation/filtration[c]	Results (positive cultures): Total[e]	Remarks	Ref.
Lysis filtration	0.45 μm; 47 mm; M	0.025% Triton X-100, 0.04% Na_2CO_3; 100 ml	3 ml	336 (173)	196 (33)	369 (2)	Solution used for lysis was toxic for many bacterial species; toxicity was greatest for Gram-negative bacteria and least for enterococci and staphylococci; filter clogging was a problem and was accentuated when blood, anticoagulated with SPS, was allowed to stand prior to lysis; data for broth comparison are from a single aerobic bottle	21
Lysis filtration	0.45 μm; 90 mm; M	0.005% Triton X-100, 0.08% Na_2CO_3; 190 ml	5 ml	24 (8)	23 (8)	77 (5)	Total positives include data from two anaerobic broth methods, one of which detected 20 positive cultures exclusively; data for broth comparison are from a single aerobic bottle; filtered material washed with saline	23
Lysis filtration	0.65 μm; 25 mm; NS	0.004% Triton X-100, Rhozyme 41®, 37.5 ml	3 ml	Not done			Simulated blood cultures studied using radiometric detection; toxicity of alkaline Triton X-100 demonstrated; Triton X-100 and Rhozyme 41 without Na_2CO_3 shown to be noninhibitory for the organisms tested	27
Lysis filtration	0.45 μm; 90 mm; M; or 0.45 μm; 47 mm, three; G	0.005% Triton X-100, 0.08% Na_2CO_3, Varidase®, pH 10.0 to 10.3; 190 ml	8.3 ml	13 (3)	22 (12)	30 (4)	Varidase, a mixture of streptokinase and streptodornase, was added to prevent filter clogging due to gelation of lysed blood with elevated leukocyte count; SPS was added as an anticoagulant for blood used in lysis procedure; lysing solution did not lead to significant loss of bacteria, unless filtration time was lengthy; filter washing was shown to be unnecessary	7, 24

TABLE 1 (continued)

Comparison of Recent Methods for Concentration of Bacteria Present in Blood

Method	Filter type (pore size; diameter; supplier)[a]	Solution used for sedimentation or lysis[b]	Effective blood volume treated	Results (positive cultures)			Remarks	Ref.
				Broth comparison[c,d]	Centrifugation/filtration[c]	Total[e]		
Lysis centrifugation	Not used	12% (wt/vol) Solryth®, 0.3 ml; sucrose-gelatin mixture, 1.5 ml	6.6 ml	64 (19)	99 (54)	135 (3)	SPS used to anticoagulate blood; contamination rate 9.3% for lysis-centrifugation and 1.8% for broth method	11, 12
Lysis filtration	0.45 μm; NS; NG	0.1% Triton X-100, 0.01 M $NaHCO_3$-Na_2CO_3 buffer, Rhozyme 41, pH 10.8; 20 ml	1.5 ml diluted to 30 ml with broth	30 (4)	49 (23)	53 (2)	Impedance changes in broth culture of filter were used to monitor bacterial growth; complete lysis gave rapid filtration; incomplete lysis occurred with too much blood, an elevated leukocyte or erythrocyte count, and blood from certain patients; toxicity of lysing solution was measured and was not considered sufficient to affect the usefulness of the technique; pH after mixing with diluted blood was 7.8; retention of bacteria by membranes with various pore sizes was measured	8, 25

generally not possible. In addition, many of the broth methods used for comparison did not incorporate an anticoagulant, nor were routine subcultures made within the first 24 hr after inoculation. They would, therefore, be considered less than optimal by comparison to procedures recommended by Washington[5] and Harkness et al.[28] Nevertheless, an attempt has been made in Table 1 to extract essentially comparable data from the published accounts. Where it has been possible, data from positive cultures due to fungi have been omitted. Fungi are relatively more resistant to lytic agents,[29] and the inclusion of data that involves a large proportion of fungal cultures might bias results from procedures that were clearly detrimental to bacteria.

The proponents of filtration and centrifugation techniques have claimed that these techniques offer several advantages over conventional methods for blood culture. In particular, filtration techniques have been shown to give better recovery of bacteria from blood of patients that are receiving antimicrobial therapy.[7,8] Sullivan et al.[24] showed that the failure of earlier studies[18,19] to clearly demonstrate this advantage was probably due to filter clogging that prevented adequate drainage or washing of the retained bacteria to remove residual antibiotics. Sullivan et al.[7] found that, of 19 positive cultures from patients receiving antibiotics, a lysis-filtration method detected 14; whereas, the comparison broth method detected only 8. Kagan et al.[8] have reported three brief case histories in which the advantage of a lysis-filtration technique for detection of bacteremia during antimicrobial therapy was illustrated. They also reported that lysis filtration was more sensitive than conventional broth methods, and that transient or subclinical bacteremias might be detected. The most recent studies (Table 1) have shown that more positive cultures were detected by lysis-centrifugation or lysis-filtration techniques than by conventional broth methods.[7,8,12] Statistical analyses of these data, however, were not reported.

Almost all of the studies that have reported new techniques for processing blood cultures by centrifugation or filtration have reported that bacteremia was detected and identified more rapidly by the new procedure than by the broth method used for comparison.[7,12,17-21,23] Finegold et al.[18] reported an average time of 28 hr for detection by a filtration method; whereas, a broth method required 61 hr, and pour plates required 46 hr. These data for the broth method may be biased, since routine subcultures were performed only before discard. Kozub et al.[17] performed routine subcultures at 24 and 72 hr and found that a filtration method allowed isolation and preliminary identification in approximately one half the time of a parallel broth method. Winn et al.[19] reported an average detection time of 17 to 19 hr for a filtration technique and 24 to 31 hr for a broth technique. In contrast to these results, Kagan et al.[8] compared a conventional broth technique with a filtration technique in which the filter was cultured in broth. They found little advantage in speed, except that which could be attributed to the continuous monitoring system that was used. Todd and Roe[30] and Harkness et al.[28] have shown that the time for detection and identification of bacteremia by conventional broth cultures was significantly decreased when routine subcultures were performed soon after receipt of a culture in the laboratory. In the experiments of Todd and Roe[30] for example, 85% of the positive cultures were detected within 24 hr. The average times for detection and identification were 22 hr and 24 hr, respectively.

The centrifugation and filtration techniques that use solid media for culture allow detection of primary growth as discrete colonies. This may have advantages compared to broth techniques that do not include an early subculture. Several studies have reported that presumptive identification could be made sooner by using rapid biochemical tests and selective or differential media.[17-19] Direct susceptibility tests have been performed using the isolated colonies,[17-19] and early recognition of polymicrobial bacteremia has been reported to be facilitated.[12,19,23] The primary growth on solid media allows quantitation of bacteremia. This may not be a significant advantage per se,[28] but it has been reported to be of help in deciding whether a particular organism was a contaminant.[12,17-19]

It may be worthwhile to note that most studies which use filtration describe a 0.45-μm pore-size filter, although larger pore sizes have been shown to be just as efficacious in retaining bacteria and less susceptible to clogging.[18,22,25,27] Reports from studies of environmental monitoring of water have shown additional advantages to the use of larger pore sizes that may be relevant to blood culture techniques. Sladek et al.[31] showed that maximum fecal coliform recoveries were obtained by using membranes with a pore size of 0.7 μm. Compared to membranes with a retention pore size of 0.45 μm, recoveries of fecal coliforms were threefold greater. It was postulated that this increase in recovery was related to better diffusion of nutrients through the larger pores, and that the larger surface openings allowed captured organisms to be buried in the membrane and protected from drying. In support of this, it was shown that when different pore-size membranes were overlayed with agar, fecal coliform recoveries were about equal. Plating the membranes with the upstream surface against the agar gave similar results, although quantitation was less precise.

In summary, the relative advantages provided by concentration techniques for more rapid detection and identification of bacteremia are dependent upon the elegance of the conventional system used for comparison. Concentration techniques are at least comparable in sensitivity to conventional methods. They may provide better recovery in some situations as a result of removal of inhibitory substances in blood. The problem of filter clogging has been largely solved, and this has allowed a larger volume of blood to be processed more rapidly.[7,8,17,24,25] Although the time for processing blood cultures by a filtration technique has decreased, comparative time studies with conventional methods have not been reported. Anaerobic bacteria have been isolated by a filtration technique,[17] but clinical experience using filtration techniques for anaerobic bacteria is limited. Lysis centrifugation is only recently reported, and its evaluation relative to filtration techniques must await further studies.

METHODS FOR RAPID DETECTION OF BACTERIAL GROWTH

Radiometric Detection

Radiometric detection of $^{14}CO_2$ produced by bacterial metabolism of a ^{14}C-labeled substrate was described in 1956 as a method to detect rapidly bacterial contamination of water.[32,33] Filter paper impregnated with a solution of $Ba(OH)_2$ was used to collect the $^{14}CO_2$ produced, which was then measured by conventional equipment for detection of radioactivity. Levin et al.[33] used [1-^{14}C] lactose as a substrate and were able to detect low numbers of *Escherichia coli* in less than 1 hr. Schrot et al.[27] in 1973 applied a modification of this technique to the detection of bacteria in blood. Bacteria were recovered from simulated blood cultures by using a lysis-filtration technique. The membrane filter was placed in a chamber containing a small amount (0.1 ml) of radioisotopically labeled medium. Attached to this chamber was a second chamber that contained an absorbant pad wetted with a drop of $Ba(OH)_2$ solution. The $Ba(OH)_2$ pads were changed at intervals, and the amount of $^{14}CO_2$ that had been collected was measured. This technique presumably combined the sensitivity of radiometric detection with the advantages of lysis filtration for removal of inhibitory substances in blood. Clinical studies demonstrating the advantages of this technique compared with conventional and other radiometric techniques for blood culture have not been reported. An interesting modification of this technique that would allow the continuous monitoring of the cumulative $^{14}CO_2$ produced has been described, although application to blood cultures has only been suggested.[34,35]

An alternative approach to the radiometric detection of bacteria was described by DeLand and Wagner[36] in 1969. In this approach, $^{14}CO_2$ produced by bacterial metabolism of [U-^{14}C] glucose in a broth medium was flushed into the ionization chamber of a monitor for radioactive gas. They extended this method to a study of simulated and actual blood cultures, and demonstrated that bacterial metabolism in simulated blood cultures containing between 4 and 56 organisms could be detected within a few hours and before visible evidence of growth.[37] They also reported that in a parallel study with a conventional culture method of 500 blood cultures, bacteria were detected in 30 specimens by the radiometric procedure and in only 26 by the routine method. No specimens were positive by the routine method that were not detected radiometrically, and detection times were consistently shorter by the radiometric method. Washington and Yu,[38] however, were unable to confirm these earlier results in their study of simulated blood cultures and a small number of actual blood cultures. They reported particular difficulty in detection of group D streptococci and *Pseudomonas aeruginosa.* Waters[39] later showed with broth cultures containing only ^{14}C-glucose as a source of $^{14}CO_2$ that detection of *P. aeruginosa* and *Streptococcus pyogenes* was delayed com-

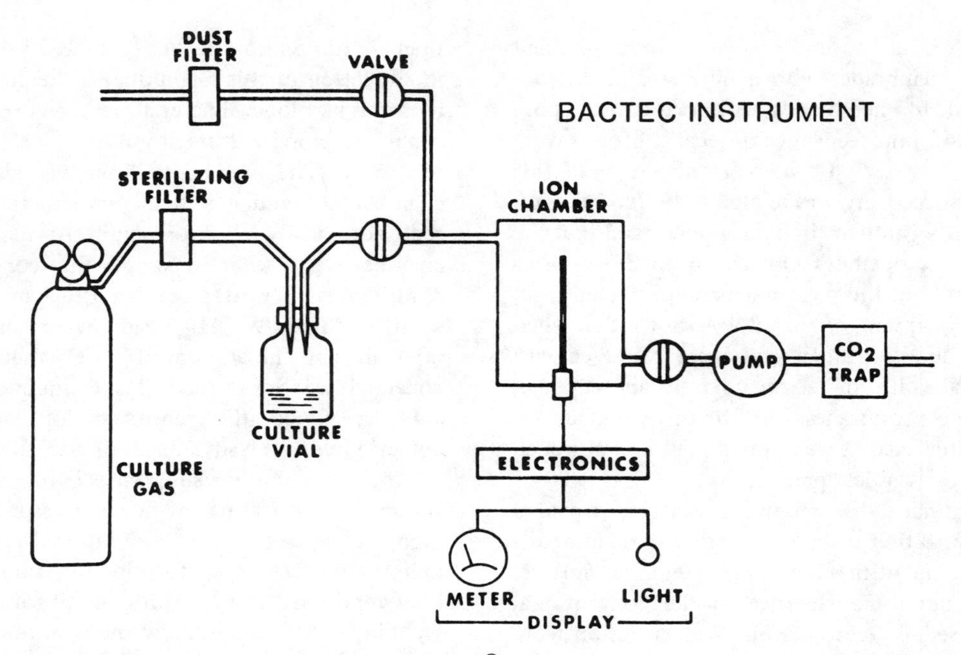

FIGURE 1. Diagram of the basic BACTEC® instrument. (Courtesy of John R. Waters, Johnston Laboratories, Cockeysville, Md.)

pared to several species of Enterobactericeae and *Staphylococcus aureus.*

The BACTEC® System

The BACTEC system (Johnston Laboratories, Cockeysville, Maryland) is a commercially available system that can be used for the radiometric detection of microorganisms in blood. This system is based on the original concept of DeLand and Wagner;[36] however, several improvements have been made, including the use of a proprietary mixture of ^{14}C-labeled substrates in the media. A diagram of the basic instrument is shown in Figure 1. A culture vial that is sealed with a rubber septum is placed on the instrument for testing. Two sterile needles pierce the septum and penetrate into the gas space above the medium. The appropriate gas to be used for incubation is drawn through a sterilizing filter, through the culture vial, and into the ionization chamber. In this way, $^{14}CO_2$ that has been produced by metabolism of various ^{14}C-labeled substrates and released into the headspace above the broth medium is flushed into the ionization chamber and measured. This measurement is converted to a growth index reading (GI), which is based on an arbitrary linear scale related to the amount of radioactivity in the ionization chamber. A reading of 100 corresponds to 0.025 μCi of carbon-14. The ionization chamber is exhausted through a CO_2 trap to prevent release of radioactivity into room air, and sampling needles are automatically heat-sterilized between each testing cycle. Three models of instrument are available for routine applications. These models vary primarily in the extent of automation and the number of vials that may be loaded onto the instrument at one time. The model 301 is described as a semiautomated instrument. An operator is needed to place each culture vial on the instrument for testing. The model 225 is described as fully automated. A rack of 25 culture vials may be placed on the instrument at one time. Testing is sequential, and results are printed on paper tape. A threshold GI may be set, and the positions of cultures that equal or exceed this limit are flagged. The description of the model 460 is similar to that of the model 225, except that 60 culture vials may be loaded onto the instrument at one time.

Evaluation of the BACTEC System

The BACTEC system is in widespread use in more than a hundred clinical laboratories.[185] Proponents of the system have claimed that it offers advantages for early detection of septicemia with sensitivity comparable to conventional methods[40-47] and that the automated models decrease laboratory workload for the processing of blood cultures.[43] (A recent review of the BACTEC system has been published.[48])

Any comparison of two systems for blood culture is markedly influenced by the adequacy of each system. The comparison of the BACTEC system with conventional blood culture systems is further complicated by the fact that both systems are in a state of evolution and improvement. An optimal system for conventional blood culture or radiometric detection has not been defined, although one can point to obvious deficiencies in both systems in various studies. The conventional and radiometric methods used in different comparison studies are outlined in Table 2. Before considering the results from these comparison studies, however, it is necessary to consider studies in which particular variables in the radiometric and conventional methods were analysed. In this way, one may evaluate the relative efficiencies of the systems that were used for comparison.

Agitation of BACTEC Cultures

DeLand and Wagner[37] showed that evolution of $^{14}CO_2$ from an aqueous solution could be increased by agitation and by decreasing the pH. Adjustment of pH was not practical for bacterial cultures, because many organisms were inhibited by an acidic medium.[36,37] Agitation, however, did not adversely affect bacterial growth and did lead to more rapid detection in some instances.[36,37,49,50] For cultures of *P. aeruginosa,* agitation has been shown to give only more rapid detection.[42,50] For several other aerobic organisms, agitation gave a higher peak GI, or a higher peak GI and faster detection.[49,50] Agitation had no effect on the detection of aerobic or anaerobic organisms grown in prereduced media under anaerobic conditions.[49,50]

The recommendation of the BACTEC instrument manufacturer is that aerobic test vials should be agitated for at least the first 24 hr of incubation. This recommendation applies to both the isotonic 6A medium as well as the hypertonic 8A medium.[51] Flasks containing both of these media are provided with stirring magnets. Flasks containing prereduced media for anaerobic cultures (7A or 7B) do not contain stirring magnets.

In many of the experiments described in Table 2, agitation of cultures during incubation was not explicitly described. In each of these instances, the BACTEC model 225, which incorporates provision for agitation of cultures, was used. It is likely that differences in agitation of aerobic cultures were not a major variable in these studies.

Agitation of Conventional Cultures

Ellner et al.[52] have recently shown that agitation of conventional aerobic cultures can give better recoveries of various organisms, including yeasts. The conventional systems used in comparison studies (Table 2) did not utilize agitation.

BACTEC Media and Atmosphere of Incubation

The media used in original development of radiometric detection contained [U-^{14}C] glucose as the sole source of $^{14}CO_2$, and air was used for incubation.[36–39,42] These conditions resulted in poor detection of microaerophilic streptococci,[42] anaerobic bacteria,[38] and nonfermentative bacteria such as *P. diminuta* and *Alcaligenes faecalis.*[39,53] A mixture of ^{14}C-labeled substrates was developed that allowed detection of organisms that do not produce detectable quantities of CO_2 from glucose.[53] The composition of this mixture is proprietary, although Caslow et al.[54] reported that the mixture contained ^{14}C-labeled glucose, as well as simple alcoholic, carboxylic, and amino compounds. Previte et al.[55] have reported a nonproprietary medium for the detection of nonfermentative bacteria. The ^{14}C-labeled components of this medium were [U-^{14}C] glucose, [5-^{14}C] glutamate, and [^{14}C] formate. The total radioactivity of this medium per 30 ml of broth was 3.75 μCi. In contrast, the BACTEC medium contains only 1.5 μCi per 30 ml of broth. Earlier detection of *A. faecalis* was observed compared to the BACTEC medium, but this may have been related to the higher activity.

Different BACTEC media consist of 1.5 μCi of the ^{14}C-labeled substrate mixture in 30 ml of media. The basal medium in each contains tryptic digest of casein soy broth (TSB), hemin, menadione, and sodium polyanetholsulfonate (SPS). The aerobic medium (6A) contains the basal medium with ^{14}C-labeled substrate and a CO_2 atmosphere. The anaerobic medium (7A) is prereduced and contains the basal medium with ^{14}C-labeled substrate, L-cysteine, yeast extract, and an atmosphere of nitrogen and CO_2. An improved anaerobic medium (7B) has been introduced. The hypertonic medium (8A) has the same composition as 6A medium with the addition of 10% sucrose. An early study[40] used a hypertonic medium (4A), the description of which was identical to that of 8A medium.

TABLE 2

Description of Various BACTEC and Conventional Culture Systems[a]

BACTEC System						Conventional System						
						Media						
Media	Blood volume (ml)	Culture atmosphere	Threshold GI	Routine subculture[b]	Macroscopic inspection	Type	Volume (ml)	Blood volume (ml)	Culture atmosphere	Routine subculture[b]	Comment	Year and ref.
Special, 30 ml	2	air	20	ND	NS	TSB	70	8	NS	12–18 hr, 10 days		1971[42]
						TG	20	2	NS	12–18 hr, 10 days		
6A	3	NS	30	14 days	NS	TSB	100	10	CO_2, NV	≤24 hr		1973[69]
7A	3	NS	30	14 days	NS	Thiol®	100	10	CO_2, NV	≤24 hr		
6A	3–5	10% CO_2	20	ND	NS	CB	50	5	air, V	10 days	Gram stain only was done at 72 hr	1974[41]
6A (?)	3	10% CO_2	30	14 days (?)	NS	VCT	18	2	10% CO_2, V	12–18 hr	Additional Gram stain after 5 days	1974[45]
7A (?)	3	10% CO_2; 85% N_2; 5% H_2	30	14 days (?)	NS							
6A	2.5	10% CO_2	NS	7 days	NS	CB	(NS	2.5	10% CO_2	1, 7, 14 days	Venting of conventional culture was not described	1975[43]
7A	2.5	10% CO_2; 85% N_2; 5% H_2	NS	14 days	NS							
6A	2–3	10% CO_2	30	5 days	daily	TSB	50	4–5	CO_2, NV	2, 7, 12 days		1975[47]
7A	2–3	10% CO_2; 85% N_2; 5% H_2	30	5 days	daily	TG	50	4–5	CO_2, NV	2, 7, 12 days		
6A	3	NS	30	7 days	NS	TSB	50	5	CO_2, NV	12 to 18 hr, 14 days	GI ≥ 20 increasing to ≥ 30 was considered positive	1975[44]
7A	3	NS	30	7 days	NS	Thiol®	50	5	CO_2, NV	12 to 18 hr, 7, 14 days		
6A	3	5% CO_2	30	7 days	daily (?)	ND	ND	ND	ND	ND		1977[73]
7A	3	10% CO_2; 90% N_2	30	7 days	daily (?)	ND	ND	ND	ND	ND		
ND	ND	ND	ND	ND	ND	TSB	100	10	CO_2 V	3 to 19.5 hr, 1, 5 days	Examined macroscopically during day of receipt, daily for 7 days, and after 14 days	1975[28]
ND	ND	ND	ND	ND	ND	TSB	100	10	CO_2, NV	3 to 19.5 hr, 1, 5 days		

[a]Abbreviations: ND, not done; NS, not specified; NV, not vented; V, vented; VCT, Vacutainer® Culture Tube, Becton-Dickinson, Rutherford, N.J.; CB, Columbia broth; TG, thioglycollate broth; TSB, tryptic digest of casein soy broth; Thiol®, Thiol broth, Difco.
[b]Time in days or hours after obtaining blood or inoculation of culture.

An early criticism of BACTEC 6A medium was based on the absence of pyridine nucleotide (V factor) in this medium. Larson et al.[56] were unable to detect species of *Haemophilus* that require V factor for growth when the 6A medium was inoculated with pure cultures of these organisms. Subsequent studies[40,45] have detected *Haemophilus influenzae* from blood cultures using the 6A medium. The absence of V factor in BACTEC media is possibly of significance only when these media are used for detection of bacteria in fluids other than blood. Rosner, however, has reported[57] radiometric detection of *Haemophilus* species in pleural fluid, synovial fluid, and cerebrospinal fluid (CSF), using the hypertonic 8A medium without added V factor.

Zwarun[58] studied the effect of osmotic stabilizers on the production of $^{14}CO_2$ by bacteria and blood. BACTEC 6A medium was used for comparison and as a basal medium to which different osmotically active agents were added. Of the agents tested, he concluded that 10% sucrose was potentially the most useful. This agent did not retard growth of *Streptococcus pneumoniae, Haemophilus* spp., *P. aeruginosa, P. diminuta,* or *S. pyogenes;* and although bacterial production of $^{14}CO_2$ was lowered, production of $^{14}CO_2$ by blood was also lowered. Bannatyne and Harnett[40] compared BACTEC 6A medium with a hypertonic medium containing 10% sucrose for detection of bacteremia in neonates. Although parallel cultures could not be performed, they observed no false-positive blood cultures with the hypertonic medium. By comparison, the 6A medium gave 67% false-positive cultures with GIs greater than 20 and 15% with GIs greater than 30. The atmosphere of incubation was not reported for either medium.

Coleman et al.[59] compared aerobic (6A), anaerobic (7A), and hypertonic (8A) media for detection of bacteremia in 5811 blood cultures. Parallel procedures were used with threshold GI value of 30 for aerobic and anaerobic vials and 20 for the hypertonic vials. The atmosphere for aerobic and hypertonic vials was 10% CO_2 and 90% air. For anaerobic vials, a mixture of 80% nitrogen, 10% CO_2, and 10% hydrogen was used. Excluding probable contaminants, 80 cultures were positive only in the hypertonic medium, compared to 43 and 40 cultures positive only in the aerobic or anaerobic media respectively. Septicemia was detected in 57 patients with only the hypertonic medium compared to 37 patients with only the aerobic medium ($p < 0.025$). The major difference was in detection of Gram-positive cocci. It was not determined whether this difference could be related to antibiotic therapy. The hypertonic medium detected fewer false-positive cultures (1.7%) than the aerobic medium (19%); however, the percentage of false-positive aerobic cultures was reduced to 6% when 5% CO_2 rather than 10% CO_2 was used to flush the vials. Detection times for the two media were not significantly different. These authors concluded that all three media should be used routinely.

Caslow et al.[54] compared aerobic, anaerobic, and hypertonic media containing 10% sucrose. These media were prepared by the addition of 1.5 μCi of the ^{14}C-labeled substrate used in the BACTEC media to a modified Columbia broth. Additional glucose (0.25%) was added to the anaerobic medium. These media were inoculated in parallel. The atmosphere for aerobic and hypertonic media was 10% CO_2 in air. For anaerobic cultures, a mixture of 85% nitrogen, 10% CO_2, and 5% hydrogen was used. The threshold GI values for aerobic and anaerobic cultures were 35 and 30, respectively. For the hypertonic medium, a threshold GI of 20 was used for readings at 24 and 48 hr, and a value of 35 was used for readings after 7 days. A total of 1000 cultures were tested, from which 104 isolates were recovered. After 24 hr of incubation, 73 aerobic cultures gave positive GIs of which 26 (36%) were considered false positives. False-positive readings were obtained from only 1 (5%) of 20 positive anaerobic cultures and 8 (12%) of 65 positive hypertonic cultures. An additional 127 aerobic cultures developed positive GIs between 2 and 7 days of incubation; 116 (91%) of these were false positives. In contrast, only 23 hypertonic vials developed positive readings between days 2 and 7, of which 15 (65%) were false positives. The authors suggested that the threshold for aerobic cultures could be increased to 45 at day 7. This would have greatly reduced the number of false-positive vials without affecting the detection of positive cultures. Although 22 isolates were recovered only from the hypertonic medium, the authors were unable to determine whether this was due to hypertonicity of the medium or to random variation.

Hull et al.[60] compared BACTEC aerobic (6A), anaerobic (7A or 7B), and hypertonic (8A) media for recovery of organisms from 2750 blood cultures. There was no statistically significant difference between the 6A and 8A media, except that presumed contaminants were isolated more frequently from the 8A medium ($p < 0.05$). They also found that using all three media (6A, 7A or 7B, 8A) gave no statistically significant advantage in recovery of organisms compared to combinations of 6A or 8A media with 7A or 7B media. Kulas et al.[61] observed decreased radiometric detection of pneumococci in an osmotically stabilized medium, presumably due to the high acidity produced by pneumococci in that medium. The manufacturer of BACTEC media has reported[51] unpublished data showing that hypertonic 8A medium may have advantages for detection of bacteremia in the presence of antibiotics, but that recovery of pneumococci may be reduced. It is also claimed[51] that the number of false positives is reduced and that a threshold GI of 20 should be used.

There are conflicting reports also on the role of hypertonic media in conventional blood cultures.[5] In recent studies, Rosner[62] has found that a hypertonic medium containing 10% sucrose resulted in more isolates compared to the identical medium without sucrose; however, survival of *S. pneumoniae, Neisseria meningitidis,* and *H. influenzae* was shorter in the hypertonic medium. Statistical analysis of these differences was not reported. Washington et al.[63] found no advantage in using a medium containing 15% sucrose over an identical isotonic medium. *Bacillus* was isolated more frequently in the hypertonic medium ($p < 0.01$); whereas, *Haemophilus, Staphylococcus aureus,* and Bacteroidaceae were isolated more frequently ($p < 0.05$) from the medium without sucrose. Ellner et al.[52] concluded that osmotically stabilized media may be useful in detection of relatively rare cases of endocarditis or septicemia involving organisms that fail to grow in conventional isotonic broth. They also found that *Pseudomonas* was isolated more frequently from the isotonic medium under stationary aerobic conditions, which was attributed possibly to decreased oxygen tension in the hypertonic medium. Rosner[64] and Washington et al.[63] have commented that macroscopic examination of hypertonic media is complicated by the cloudiness and hemolysis that normally occurs after incubation.

The currently recommended culture gas for BACTEC aerobic 6A and hypertonic 8A media is 5% CO_2 in air.[51,65] The original comparison study[42] used only air. Waters and Zwarun[53] subsequently showed that 10% CO_2 in air enhanced detection of streptococci, *Haemophilus,* pneumococci, and pseudomonads. The mixture with 10% CO_2 gave higher background production of $^{14}CO_2$ from blood and necessitated increasing the threshold GI to 30 for aerobic 6A medium. Under these conditions, an increased number of false-positive cultures still occurred;[45,59] however, this number was greatly reduced by decreasing the CO_2 concentration to 5%.[59] False-positive cultures due to elevated leukocyte counts have also been reported;[47,66] however, Smith and Little[45] found no common denominator in the clinical histories of the patients in their study from whom false-positive cultures were obtained. Bannatyne and Harnett[40] have also shown that neonatal blood may give high $^{14}CO_2$ production, which was suppressed in a hypertonic medium.

Macroscopic Examination of BACTEC Cultures

Macroscopic inspection of BACTEC media has been shown to be necessary in order to obtain optimal detection of positive cultures.[66-68] This is of particular importance when using the hypertonic 8A medium, since gas pressure may be generated from fermentation of sucrose. Testing of these samples can lead to bacterial contamination of the BACTEC instrument.[51] Wood[68] found that 75% of the anaerobic 7B vials tested failed to register a GI of ⩾25, but were positive by visual inspection. Unfortunately, the number of anaerobic cultures tested was small, and the medium had expired just prior to the beginning of this study. With the single exception of the study by Thiemke and Wicher,[47] comparison studies (Table 2) have not specified whether macroscopic inspection of BACTEC culture vials was performed.

Frequency of Radiometric Testing

The frequency of radiometric testing can affect the detection time. Washington and Yu[38] tested cultures several times in the first 48 hr after receipt; however, no testing was done between 6 and 18 hr. This factor was used by DeBlanc et al.[42] to explain apparent discrepancies, since in their work hourly testing was performed. Anaerobic organisms produce CO_2 more slowly; therefore, anaerobic culture vials are usually tested less frequently than aerobic vials. Randall[44] has reported that for anaerobic cultures, a rising GI that is less than 30 may still be indicative of a positive culture. Most studies have reported testing aerobic and hypertonic cultures multiple times

TABLE 3

Detection of Bacteremia by BACTEC and Conventional Culture Systems[a]

Number of cultures	Results (positive cultures or isolates)[b] System[a]	Each system	Total	False positives[c]	Comment	Ref.
2967	BAC	111 (13)	138	None	Numbers do not add up properly. Data based on number of cultures	42
	CON	125 (23)				
1445	BAC	87 (22)	111	175 (12)	Conventional method failed to detect 8 of 9 *Pseudomonas* spp. Data based on number of isolates	69
	CON	89 (24)				
1261	BAC	283 (38)	311	545 (43)	Many of the cultures were from patients in a burn unit. More than one isolate from 30% of the cultures. Data based on number of cultures. A threshold GI of 30 would have reduced false positives to 19 (1.5%)	41
	CON	273 (28)				
1194	BAC	97 (47)	105	106 (8.9)	Data based on number of isolates	45
	CON	58 (8)				
1121	BAC	100 (15)	108	NS	Ten cultures positive only in 7A medium not included in results. Data based on number of cultures	43
	CON	93 (8)				
3045	BAC	214 (45)	280	208 (6.8)	Multiple isolates from 18 cultures. BACTEC data do not include 14 false-negative cultures. Data based on number of isolates	47
	CON	235 (66)				
1530	BAC	121 (32)	140	24 (1.6)	BACTEC data do not include 4 false-negative cultures. Data based on number of isolates	44
	CON	108 (19)				

[a]Abbreviations: BAC, BACTEC; CON, Conventional.
[b]Data do not include presumed contaminants.
[c]Number in parentheses is the percentage false positives of the total number of cultures; NS, not specified.

during the first 24 to 48 hr of incubation, while anaerobic cultures are tested no sooner than 24 hr after receipt.

Specificity of Radiometric Detection

The number and percentage of radiometrically false-positive cultures that have been reported in various comparison studies are shown in Table 3. These false-positive cultures are significant from the standpoint of increasing laboratory workload. False-positive cultures have occurred most frequently with the aerobic 6A culture medium. The percentage is related to the concentration of CO_2 in the incubation gas and inversely related to the threshold GI that is chosen. Coleman et al.[59] found that using 5% CO_2 instead of 10% decreased their false positives from 19% to 6% of the total aerobic cultures. DeBlanc et al.[42] and Smith and Little[46] used air as a culture gas and observed a relatively low false-positive rate. The use of air as a culture gas, however, reduced the recovery of some organisms.[53] Brooks and Sodeman[41] observed a false-positive rate of 43% using a threshold GI of 20, but this would have decreased to 1.5% if a GI of 30 had been used. This higher threshold, however, would have prolonged detection times for *Candida.* French et al.[66] observed a 9.7% false-positive rate using a threshold GI of 20. Caslow et al.[54] used 10% CO_2 in air as a culture gas and a threshold GI of 35 for the 6A medium. They observed a false-positive rate of 14%, but felt this could have been decreased significantly without decreasing detection of positive cultures by increasing the threshold GI to 45 for the readings after 7 days of incubation. In contrast, Randall[44] reported a false-positive rate of only 1.6% using a threshold GI of 30. The gas used for culture was not specified.

Conventional Culture Media and Atmosphere of Incubation

Washington,[5] in reviewing blood culture techniques, concluded that at least two bottles, each containing 50 to 100 ml of two different media should be inoculated. Commercial media that are bottled under vacuum with CO_2 were reported to be suitable for anaerobes if unvented. He also recommended that cultures should be inspected daily and at least once on the day of receipt. Routine blind subcultures were recommended within 24 hr after inoculation and again after 4 or 5 days. Routine anaerobic subcultures were not recommended. Recent data have shown that *Pseudomonas* and *Candida* were recovered more frequently and in less time from a vented than a nonvented culture bottle.[28] The conventional procedure used in many of the comparison studies (Table 2) was deficient compared to these recommendations by Washington[5] and Harkness et al.[18] Renner et al.[69] did only one subculture and did not vent for an aerobic culture. Brooks and Sodeman[41] used a single bottle, which was incubated aerobically; however, their study did not include anaerobic incubation with either method. Routine microscopic examination of cultures was not done until after 72 hr, and a subculture was not done until day 10. This procedure would be expected to result in delayed detection of some organisms by the conventional system.[70] Smith and Little[45] used as a comparison system the Vacutainer® Culture Tube (VCT) (Becton-Dickinson, Rutherford, NJ). Hall et al.[71] compared a two-bottle VCT system with a two-bottle conventional system that used both TSB and thiol. The VCT system isolated many organisms significantly less frequently ($p < 0.01$) than the conventional system. This difference was attributed mostly to the difference in blood volume cultured. Morello[43] used a single aerobic broth bottle for comparison. Although the volume of blood used in each system was the same (2.5 ml), this volume was less than that recommended by Washington[5] for conventional systems. Thiemke and Wicher[47] did not routinely subculture until day 2 nor were cultures vented.

Comparison Studies – Sensitivity

The numbers of organisms recovered in parallel studies using the BACTEC and conventional system are presented in Table 3. Methodology for each study is described in Table 2. The data in Table 3 include recoveries of both bacteria and fungi. With the exception of the study by Smith and Little,[45] there was comparable recovery by both systems. In no case was statistical analysis of the differences reported with regard to the total number of isolates or the isolation of specific organisms. Some observations regarding specific organisms, however, were made. Renner et al.[69] reported that detection efficiency was about equal for the two systems used, except BACTEC detected more *Torulopsis glabrata* and *Pseudomonas* spp., and the conventional system detected more anaerobes and *Enterobacter aerogenes.* Brooks and Sodeman[41] studied only aerobic isolates and found that the BACTEC system detected more *Candida* spp. and *Streptococcus faecalis.* Thiemke and Wicher[47] found essentially equal recoveries, except the conventional system recovered more anaerobes. They felt that this difference could be only partially explained by the difference in volume of blood used in each system. They did not evaluate the newer 7B anaerobic medium. Randall[44] observed better detection of yeasts with the BACTEC system; however, in her study, the BACTEC system also recovered more anaerobes, and overall recovery by the BACTEC system was better. Renner et al.[69] claimed to have done statistical comparisons, but did not report the findings from these analyses. It might be assumed that these analyses failed to show any significant differences in detection efficiency.

Of great significance, is the number of BACTEC cultures that fail to show a positive GI, but are found to contain viable organisms when subcultured. These are termed false-negative cultures and should not be affected by chance variation as a result of inoculation with a small volume of blood that possibly contains a low concentration of bacteria. The volume of blood used for inoculation has been shown to affect primarily the recovery of Gram-negative bacilli,[72] and the BACTEC system has generally been found suitable for detection of these organisms. The identity of organisms that were found to give false-negative BACTEC results are shown in Table 4. Studies by Renner et al.,[69] Brooks and Sodeman,[41] and Morello[43] either did not perform blind subcultures of BACTEC vials or did not report separately organisms isolated only by this subculture technique. These studies are omitted from Table 4. French et al.[66] reported that of 402 positive cultures by radiometric, macroscopic, or routine subculturing, 8% had a GI of between 20 and 29. Caslow et al.[54] reported no false-negative cultures. In their study, BACTEC media were not used, and in 20% of the positive cultures, subcultures performed at 24 hr were positive while the BACTEC did not register positive readings until 48 hr. From the data presented in Table 4, it

TABLE 4

False-negative Results with the BACTEC System[a]

Number of false negatives[b]	Organism	Comment	Ref.
8 (8)	Microaerophilic alpha-hemolytic *Streptococcus*	Microaerophilic streptococci were all from one patient	42
1 (3)	*Neisseria gonorrhoeae*		
1 (35)	*Streptococcus pneumoniae*		
2 (4)	*Staphylococcus aureus* and beta-hemolytic *Streptotoccus*		
4 (5)	Enterococcus	Enterococci were all from one patient	46
1 (8)	*Candida albicans*		
1 (1)	beta-hemolytic *Streptococcus*		
1 (9)	*Streptococcus*		47
1 (19)	*S. pneumoniae*		
1 (38)	*Staphylococcus aureus*		
2 (103)	Gram-negative bacilli		
9 (40)	fungi		
3 (6)	*Peptostreptococcus*	Author observed false-negative results from the BACTEC 7A medium usually	44
1 (18)	*Streptococcus*		
20 (44)	*Streptococcus*, group D	Twelve of 17 isolates of enterococci not detected radiometrically were from a single patient	73
10 (269)	other organisms		

[a]An organism recovered from a radiometric culture, but not detected radiometrically is considered a false negative.
[b]Number in parentheses is the total number of cultures (positives plus false-negatives) eventually detected by subculture of the radiometric media.

appears that one blind subculture of BACTEC vials is required to detect reliably all organisms present, particularly streptococci. It was noted by Strauss et al.[73] that false-negative cultures may be strain specific; of the 17 isolates of enterococci not detected by the BACTEC, 12 were from a single patient. The most appropriate time for this subculture has not been settled. An increase in the amount of radioactive substrate from 1.5 μCi to 2.0 μCi in the 6A and 8A media may decrease the number of false-negative cultures,[185] but the effect of this change on detection of false-positive cultures would also need to be evaluated.

Comparison Studies – Detection Time

Comparative detection times reported in various studies are shown in Table 5. A small source of error occurs because of different definitions for the starting time. Nevertheless, it is apparent in most studies that the radiometric procedure detected more positive cultures during the first day of incubation. This may be in part a result of timing of subcultures and microscopic examination of the conventional cultures. In recent studies, the advantage of an early subculture to decrease detection time has been shown.[28,30] Harkness et al.[28] were able to detect 48% of positive cultures by subculture after an average of 9.5 hr of incubation. By 48 hr, they detected 99% of all positive cultures by a combination of subculture and macroscopic examination. Todd et al.[30] reported detection of 85% of all positive blood cultures within 24 hr of initial incubation when subculture after an average of 8 hr and macroscopic examination at 12 and 24 hr of incubation were done.

Summary

The BACTEC system is an alternative to conventional blood culture systems. It represents a distinct advantage over some. When a single blind subculture is included, the BACTEC system is at least comparable in terms of sensitivity to conventional blood culture systems that are acceptable by current practices. The BACTEC system may offer an advantage in detection time compared to conventional systems that do not entail early blind subculture and macroscopic examination. The BACTEC system advantage may, therefore, be due to repetitive monitoring rather than an inherent increased sensitivity. This is a signifi-

TABLE 5

Detection Times for the BACTEC and Conventional Culture Systems[a]

System	Cumulative % detected by day					% Detected first by each system			Comment	Ref.
	0.5	1	2	3	≥4	BACTEC	Conventional	Both equal		
BAC	NR	65	NR	NR	NR	70	6	24	Data based on number of cultures	42
CON	NR	4	NR	NR	NR					
BAC	NR	23	65	77	100	43	26	30	Data based on number of isolates	69
CON	NR	7.6	58	82	100					
BAC	36	NR	86	94	100	NR	NR	NR	BACTEC detected positive cultures significantly sooner ($p < 0.001$)	41
CON	NR	1.2	58	71	100					
BAC	9	65	85	NR	100	19	20	61	Cumulative % from an additional study of 369 isolates in which a conventional system was not used. Other data based on number of cultures	43
CON	ND	ND	ND	ND	ND					
BAC	37	62	76	87	100	NR	NR	NR	Data based on number of isolates	47
CON	6	40	64	79	100					
BAC	ND	ND	ND	ND	ND	NA	NA	NA	Data based on number of cultures	28
CON	NR	53	99	NR	100					

[a]Abbreviations: NR, not reported; NA, not applicable; BAC, BACTEC; CON, Conventional; ND, not done.

cant consideration in an evaluation of the BACTEC system, since other systems of automated monitoring are becoming available. The BACTEC system may eliminate two of the three blind subcultures advocated by users[28] of early subculture techniques, while providing comparable sensitivity and detection time. This may result in decreased laboratory workload. This decrease, however, may be partially offset by the clinical advantages for rapid identification, susceptibility testing, and detection of polymicrobic bacteremia that growth on a subculture plate provides.

The manufacturer has submitted to the College of American Pathologists a list of claims for the BACTEC 225 that have been verified by the Product Evaluation Committee of the Commission on Standards.[65] This information may serve a useful purpose in the evaluation of certain technical and operational aspects of the instrument. They should not be used as a basis of comparison with conventional systems, since patient populations vary. In particular, the claims do not verify any advantage of the BACTEC 225 system over conventional systems.

A principal question in evaluating BACTEC and conventional systems of comparable sensitivity and detection time is the relative technical time required and cost of each system. There are no accurate and complete studies available that answer these questions.

Electrical Impedance Detection

Interest in the relationship of the electrical conductivity of a culture medium to the metabolism of bacteria in that medium can be dated back to Stewart[74] in 1898. Stewart found that the increase in electrical conductance that occurred with bacterial growth correlated with a depression in the freezing point of the medium. He concluded that both processes were due to electrolytes produced by bacterial metabolism of various substrates. This phenomenon was observed with bacterial growth in blood and serum as well as in artificial media. In fact, Stewart speculated that differences in the magnitude and rate of change in conductivity could be used to differentiate various bacteria. Further historical developments in the use of conductivity to measure microbial metabolism have been reviewed by Hadley and Senyk.[75]

Theory

Impedance is a measure of the total opposition to the flow of alternating current in an electrical circuit. Cady[76] postulated that the impedance of a cell for measuring microbial metabolism could be described by a model consisting of a resistor and capacitor in series. The impedance of this model is related to frequency by the following equation:

$$Z = R^2 + [(2\pi fC)^{-2}]^{1/2} \quad (5.1)$$

where Z is impedance, C is capacitance, f is frequency of the alternating current, and R is resistance. By measurement of the impedance change of a culture of *Escherichia coli* in TSB at various frequencies over a period of 8 hr, he was able to show by extrapolation that the capacitative component underwent a greater change (3.33 to 5.41 μF) than the resistive component (37.5 to 27.0 Ω). This large change in capacitance was also shown to be due to the deposition of bacteria or bacterial products on the electrodes which formed a capacitor analogous to an electrolytic capacitor. The nature of this material was only briefly investigated, but it was shown in the case of *E. coli* to be both soluble and cell associated. Stoner et al.[77] have reported that contact adsorption between viable organisms and the test electrode was responsible for observed impedance changes; nonviable bacteria were not detected. Production of ion pairs from neutral molecules and reduction in size of larger charged molecules also occurs in the course of microbial metabolism and contributes to changes in impedance.[76]

A correlation between impedance changes measured by different methods and bacterial cell counts has been reported by several groups.[75,76,78] Hadley and Senyk[75] have shown (Figure 2) that impedance changes observed during growth of *E. coli* in brain heart infusion (BHI) broth parallel the growth curve determined by measurement of viable cell counts. The impedance changed exponentially with the same doubling time as predicted by direct cell counts.

The relationship of impedance (Z) to frequency (f) shown in Equation 5.1 would seem to indicate that impedance changes due to capacitance would be maximal at low frequencies. Ur and Brown[78-80] used a frequency of 10 kHz, Cady[76] used a frequency of 2 kHz, and Kagan et al.[8] reported maximal sensitivity at 2 Hz. Regardless of the frequency used, however, the approximate threshold concentration of bacteria for detection was between 10^5 and 10^6 colony forming units (CFU) per milliliter. Impedance changes and sensitivity can also be influenced by other factors. Media with high buffering capacity[78,80] or ionic strength[75] have been shown to cause a decrease in sensitivity. Electrode materials such as aluminum, nickel, nickel-boron, and copper produce erratic signals.[75] Electrodes of gold or stainless steel have been used most frequently. A paradoxical initial increase in impedance was observed by Cady with *Serratia marcescens* when tested using gold electrodes in TSB at 37°C. This effect was not seen when electrodes made of stainless steel were used. Kagan et al.[8] found that sensitivity was decreased unless electrodes were flamed to a red heat after each use. For this reason, and because of cost and strength, stainless steel electrodes were preferred. In any event, the impedance measurement must be sensitive in order to detect bacterial growth rapidly. Since temperature may cause impedance to change as much as 2% per degree centigrade, precise control of temperature during incubation and monitoring has been found to be necessary.[79,80] Ur and Brown[78,79] and Cady[76] have developed equipment that measures impedance by an electrical bridge circuit. This equipment is now available commercially. The Strattometer® developed by Ur and Brown[78] is produced by Stratton and Co., Hartfield, England. The Bactometer® developed by Cady[76] is produced by Bactomatic, Palo Alto, CA.

The impedance bridge in the Bactometer measures the ratio of the impedance of a reference sample (Z_r) to the impedance of both experimental and reference samples ($Z_e + Z_r$). Thus the observed impedance (Z_m) is

$$Z_m = \frac{Z_r}{Z_r + Z_e} \quad (5.2)$$

Changes due to temperature or other factors that occur in both samples and are proportional to Z tend to cancel out. Kagan et al.[8] have found, however, that the use of an impedance bridge may not be necessary in order to detect microbial growth.

Detection of Bacteremia

Hadley and Senyk[75] have studied the detection of bacteremia in simulated blood cultures. A Bactometer was used to measure impedance changes in a blood culture bottle compared to an uninoculated bottle which was used in the reference circuit. In the preliminary studies, printed circuit electrodes were placed horizontally in each bottle. The medium was prepared from BHI broth with yeast extract, SPS, and menadione added.

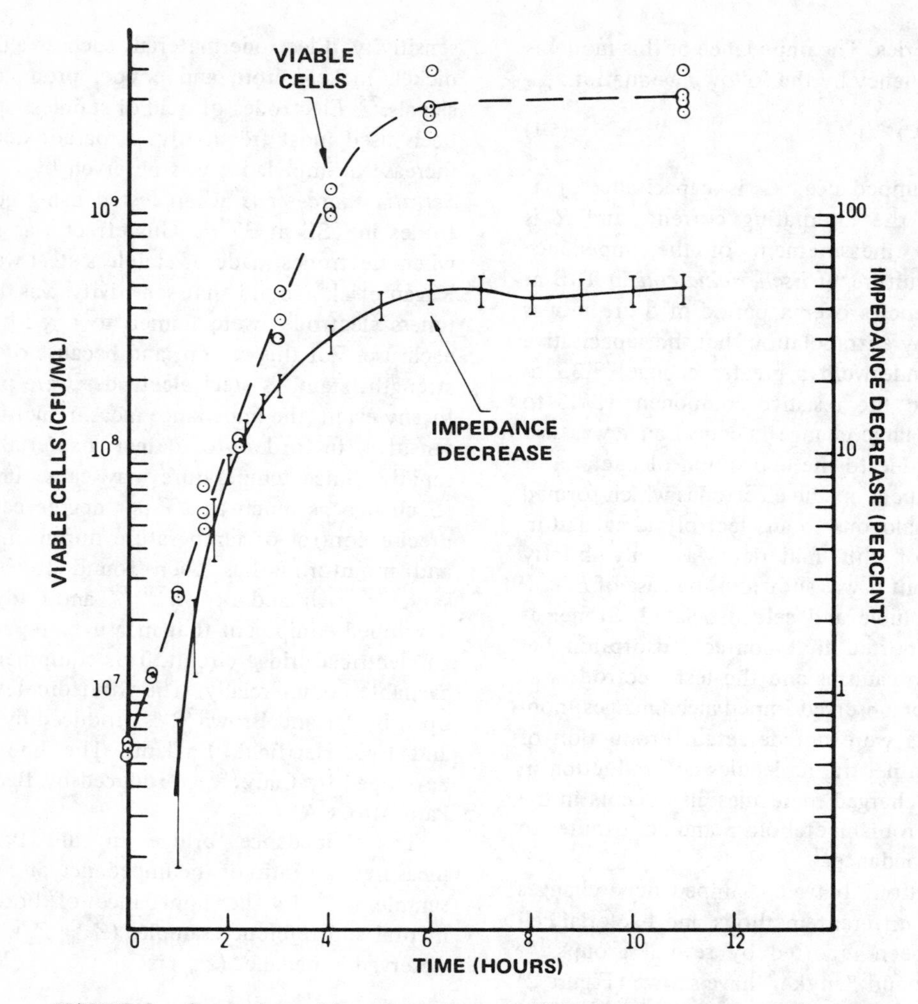

FIGURE 2. Comparison of impedance changes with bacterial growth curve. The curve, represented by a solid line, shows the impedance changes resulting from *Escherichia coli* growing in brain heart infusion broth. The data points connected by the broken line represent the result of plate counts taken from the impedance sampling chamber. (From Hadley, W. K. and Senyk, G., in *Microbiology – 1975,* Schlessinger, D., Ed., American Society for Microbiology, Washington, D.C., 1975. With permission.)

Addition of sterile blood to this medium caused an increase in impedance that was proportional to the concentration of blood. One factor that contributed to this increase was the continual settling of blood cells onto the electrodes. Addition of 0.1% agar to the medium reversed this effect. An accelerating decrease in impedance was superimposed on a background of decreasing impedance in the presence of bacterial growth. Substitution of vertical electrodes made of stainless steel resulted in lower background changes due to blood and allowed bacterial growth to be more easily detected. A comparison of the results obtained in a simulated blood culture of *P. aeruginosa* using both the horizontal (printed circuit) and vertical (stainless steel) electrodes is presented in Figure 3. Typical results obtained with simulated blood cultures and vertical electrodes are shown in Figure 4. Rapidly growing organisms were reported to be detected within 10 to 15 hr after inoculation. *Serratia marcescens, E. coli,* and *Staphylococcus aureus* were detected in 8 to 12 hr of incubation when 10^5 to 10^6 CFU/ml of bacteria were present in the culture.[82] In a later clinical study, Hadley and Kazinka[83] studied 785 blood cultures in which detection by the Bactometer and a parallel broth culture were compared. The same medium was used in each method. One

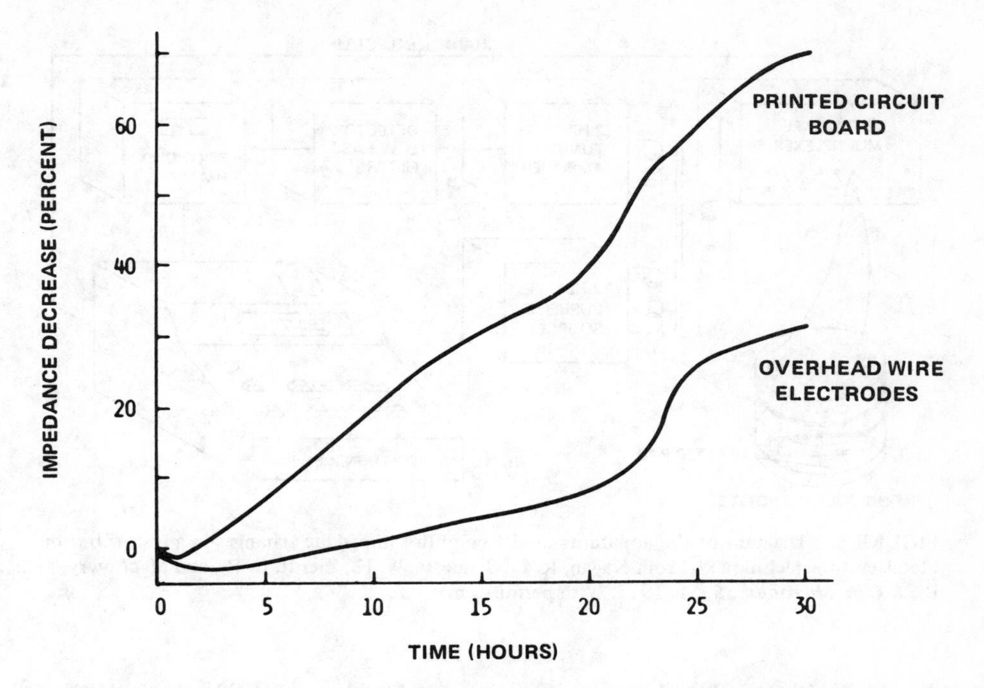

FIGURE 3. Comparison of vertical wire electrodes and horizontal (printed circuit board) electrodes in measuring the growth of *Pseudomonas* in stimulated blood culture. (From Hadley, W. K. and Senyk, G., in *Microbiology – 1975,* Schlessinger, D., Ed., American Society for Microbiology, Washington, D.C., 1975. With permission.)

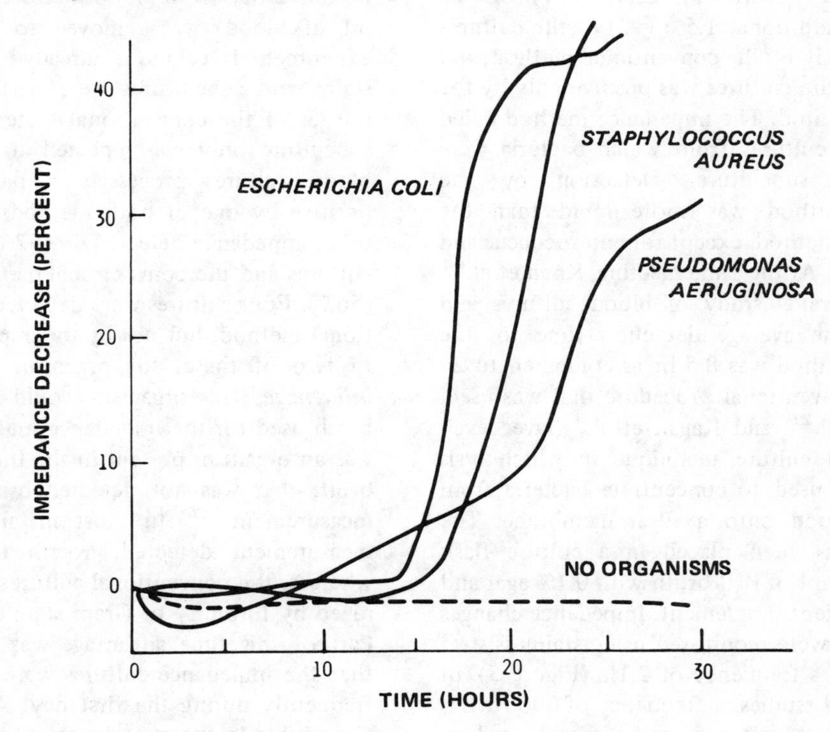

FIGURE 4. Simulated blood cultures containing 10% blood in broth. Initial culture inoculum: 1 to 10 CFU. (From Hadley, W. K. and Senyk, G., in *Microbiology – 1975,* Schlessinger, D., Ed., American Society for Microbiology, Washington, D.C., 1975. With permission.)

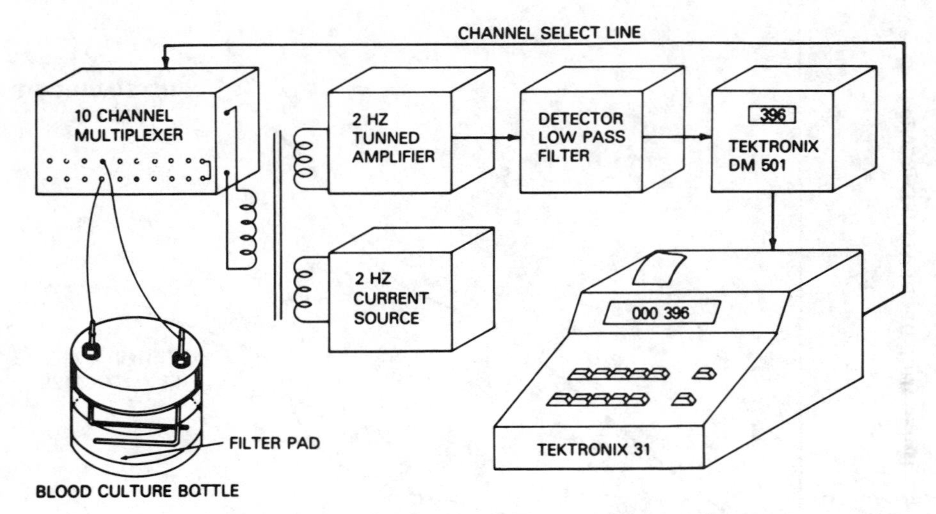

FIGURE 5. Diagram of the apparatus used to monitor impedance changes in a lysis-filtration blood culture technique. (From Kagan, R. L., Schuette, W. H., Zierdt, C. H., and MacLowry, J. D., *J. Clin. Microbiol.*, 5, 51, 1977. With permission.)

bottle inoculated with 10 ml of blood was used for each method, and the conventional method included routine subculturing and staining. Both methods were positive in 12.4% (97) of the cultures. An additional 1.5% (12) of the cultures was positive only by the conventional method, and 1.9% (15) of the cultures was positive only by the impedance method. The impedance method failed to detect 11 cultures from which bacteria were isolated on subculture. Detection by the impedance method was more rapid than the conventional method, except for enterococcus and *C. parapsilosis.* At the same meeting, Khan et al.[84] reported a parallel study of blood cultures and found that the average detection time for the impedance method was 8.5 hr as compared to 24 hr for the conventional procedure that was used.

Zierdt et al.[25] and Kagan et al.[8] have developed a blood culture technique in which lysis filtration was used to concentrate bacteria from 1.5 ml of blood onto a filter membrane. The membrane was then placed in a culture flask containing 20 ml of BHI broth with 0.1% agar and 1.5% of a nutrient supplement. Impedance changes in the culture were monitored using stainless steel electrodes and a frequency of 2 Hz (Figure 5). In more extended studies, a frequency of 800 Hz has been used.[186] A reference cell was not used. A clinical comparison of this procedure was performed with a conventional procedure in which 5 ml of blood was inoculated into each of two bottles containing 90 ml of BHI broth with SPS, *p*-aminobenzoic acid, and CO_2. One of these bottles was vented on receipt in the laboratory and 30 ml of the broth-blood mixture (containing 1.5 ml of blood) was removed to process by the experimental technique already described. Gram stains and subcultures were performed on both bottles of the conventional system at 24 hr, and subculture only was repeated at 7 days. Of 264 blood cultures processed in parallel, 53 were positive by one or both methods. The procedure using impedance detected 49 (92%) of the positive cultures and the conventional method detected 30 (56%). Four cultures were detected by the conventional method, but not by the impedance method. In two of these, the organism isolated was *H. influenzae.* This organism would not grow in the broth used for the impedance method. In no case was an organism present in the impedance culture broth that was not detected by the impedance measurement.[186] In most instances, impedance measurement detected growth in 8 to 12 hr; whereas, the conventional cultures were not recognized by turbidity or Gram stain until after 26 hr. Part of this time advantage was due to the fact that the impedance cultures were examined more frequently during the first day. A faint turbidity was visible in the impedance cultures at the same time that the change in impedance was sufficient to register as positive. The corresponding colony count at this time was 5×10^5 CFU/ml. Other advantages, probably more related to the lysis-filtration technique than to the impedance detection, were noted.

The experimental and clinical results obtained to date have shown that impedance measurement is a versatile technique for detection of microorganisms. No one classification of organism has been found that cannot consistently be detected, although the study by Hadley and Kazinka[83] would seem to present evidence that some organisms may be more difficult to detect. Septicemia due to yeasts has been reported infrequently in these studies of impedance detection. Some yeasts are known to generate an increase in impedance rather than a decrease.[75] Systems keyed to detection of only a decrease in impedance may have difficulty in detection of these organisms.

The few clinical studies to date have obtained encouraging results, but obviously more clinical studies are needed. Attention must be put on determining whether false-positive cultures are detected and on detection of cultures that are negative by impedance measurements, but from which organisms can be isolated by subculture (false-negative cultures). The advantage in detection time may be significant and may be related to the fact that impedance testing can provide almost continuous monitoring of a blood culture. More significant, with regard to providing improved detection of septicemia and decreasing laboratory workload, would be the demonstration that routine subcultures are not necessary. Development is also needed to increase the number of clinical specimens that can be processed and monitored.

Microcalorimetry

The application of microcalorimetry to the measurement of the small heat changes that occur as a result of microbial metabolism has been reviewed by Forrest.[85] More recently, Boling et al.[86] and Russell et al.[87,88] have shown that the heat changes produced by a microorganism growing in a specific medium are characteristic and may be used for identification.

Ripa et al.[89] have shown that microcalorimetry can be used to detect the growth of different microorganisms in broth media containing blood. Their experiments were designed to measure the effects of various additives, commonly used in blood culture media, on the growth of bacteria. Application to the detection of bacteremia was not proposed; however, the authors were able to detect *E. coli, S. aureus,* and *N. meningitidis* in experimentally infected blood. Blood and culture media were mixed in the proportion of 1:10. Relatively large inocula were used to give a final concentration of 10^2 CFU/ml. This technique may not be applicable to detection of bacteremia in a clinical setting, but it may be useful in the evaluation of the various media that are used for blood cultures.

Bioluminescence

The production of light from the oxidation of firefly luciferin by luciferase forms the basis of a highly specific and sensitive assay for adenosine triphosphate (ATP).[90] This assay has been applied to the detection of bacteria in aqueous suspensions,[91] as well as in clinical urine specimens.[92-94] Schrock et al.[95] have reported that the firefly luciferin assay may be used to detect bacteria growing in blood cultures. The threshold for detection of bacteria in urine specimens is generally considered to be in the order of magnitude of 10^5 organisms per milliliter,[93-94] presumably as a result of background ATP that cannot be completely eliminated.[93] Even with this threshold, studies for the detection of bacteriuria have reported significant numbers of false-positive and false-negative reactions.[92-94] Conn et al.[93] have discussed other limitations of this assay for the detection of bacteria in clinical specimens. The assay does not provide continuous monitoring. Discrete samples must be withdrawn from the culture and processed to isolate bacterial ATP before measurement can be made. Considering these factors, the procedure reported by Schrock et al.[95] does not appear promising for the clinical laboratory, particularly when compared to the simplicity of a Gram stain or subculture.

DETECTION OF BACTERIAL PRODUCTS

The methods to be described in this section do not involve the culture or isolation of a microorganism. They should be considered as adjuncts to culture rather than alternatives. Since bacterial products are detected and not bacterial growth, these methods have a theoretical advantage for detection of bacteria in situations where conventional cultures may be negative due to the presence of inhibitory substances. In many instances, these methods may provide more rapid detection, although with less sensitivity or specificity than culture. The presence of bacterial products may also indicate only a situation of previous infection, although persistence of these products has generally been found to be of short duration.

Limulus Amoebocyte Lysate Test

The amoebocyte is the cellular element in the circulatory system of the horseshoe crab (*Limulus polyphemus*). Soluble lysate of these cells forms a gel in the presence of endotoxin. Levin and Bang[96] developed a very sensitive test for endotoxin based on this fact. Rojas-Corona et al.[97] and Cooper et al.[98] compared the *Limulus* amoebocyte lysate (LAL) test with several other assay methods for endotoxin. The LAL test was far more sensitive than any of the in vitro or in vivo tests available.

Mechanism and Sensitivity

Young et al.[99] studied the mechanism of the LAL test. Amoebocyte lysate was separated into three proteinaceous fractions by using gel-filtration chromatography. Two of these fractions were shown to be involved in endotoxin-stimulated coagulation. One of these (Fraction II) contained a coagulable protein of molecular weight 27,000. This protein was consumed in coagulation and could not be demonstrated by immunodiffusion assay in the supernatant of the gel. The other fraction (Fraction I) involved in coagulation was of higher molecular weight. This fraction formed two major precipitin lines when tested by immunoelectrophoresis; however, the authors were unable to fractionate it preparatively by chromatography. Immunoelectrophoresis of gel supernatant showed that one of these components was diminished or absent, whereas the other was still present. A function for the third fraction in coagulation was not demonstrated. The endotoxin-stimulated coagulation of Fraction II by Fraction I was similar to an enzymatic reaction. A model was proposed in which a proenzyme in Fraction I was activated by endotoxin. Action of this enzyme on Fraction II gave coagulation. The rate of reaction was proportional to the concentration of endotoxin and the concentration of Fraction I. The reaction could be inhibited by treatment of Fraction I, but not Fraction II, with diisopropylfluorophosphate, parachloromercuribenzoate, and parachloromercuriphenylsulfonate. These data suggested that enzymatic activity depended upon sulfhydryl and serine hydroxyl groups.

Sensitivity of the LAL test can be influenced by the presence of inhibitory substances, the system used to detect coagulation, and the amoebocyte lysate preparation. Comparison of sensitivity between laboratories may be complicated by the failure to use a standardized endotoxin preparation or the use of endotoxin isolated from different organisms. Cooper et al.[98] were able to detect 1 ng/ml of endotoxin isolated from *E. coli* and 0.1 ng/ml of endotoxin isolated from *Klebsiella*; whereas, Yin[100] reported detection of 0.1 pg/ml of endotoxin isolated from a different strain of *E. coli*. Neither procedure used a common standard with which a comparison could be made. Two standardized preparations of endotoxin have been made available. The first was isolated from *Klebsiella*[101] and was detected in the experiments of Cooper et al.[98] at a concentration of 0.1 ng/ml. The more recent preparation was isolated from *E. coli* and was detected at a concentration of 1 ng/ml.[102] In comparison studies using the same batch of lysate, the *Klebsiella* standard described above was detected at a concentration of 6 ng/ml.[102] In most studies in which the *Klebsiella* standard was used, a sensitivity of between 0.1 and 1.0 ng/ml was reported. Quantitation of endotoxin may be based on measurement of the rate of change in turbidity,[103,104] the time for complete gelation,[98,103] or by serial twofold dilutions to obtain an endpoint.[105]

Normal human serum and plasma inhibit the activity of endotoxin in the LAL test. The inhibition is reversible, and the mechanism appears to be different from that involved in inhibition of endotoxin activity in other assays.[103] The recovery of endotoxin from serum of whole blood that is allowed to clot is less than recovery from plasma. This seems to be related to loss or sequestration of endotoxin during clotting and not to increased inhibition by serum.[103] Reinhold and Fine[105] have also recommended that heparinized plasma be used for assays, because calcium-depleting anticoagulants may activate serum esterase which can detoxify endotoxin in vitro. The nature of the inhibitor in plasma is unknown. Levin et al.[103] showed that inhibition was eliminated by simple dilution of the specimen; however, this would result in decreased sensitivity. The inhibitor was also removed by precipitation of plasma with chloroform. The resulting emulsion was centrifuged and endotoxin was recovered from a cloudy middle layer. Recovery of endotoxin activity by this procedure was essentially quantitative. Protein precipitation by heat or trichloroacetic acid resulted in complete loss of activity. Reinhold and Fine[105] developed a simple procedure based on precipitation of plasma by acidification and realkalinization. They found this procedure faster and less cumbersome than chloroform precipitation. Both methods gave good recovery of endotoxin activity, but comparison studies of the two procedures have not been reported. Hollander and Harding[104] have developed a procedure for small volumes of plasma. Inhibitory substances were removed by filtration of plasma through a column of Bio-Gel® P-200 (Bio-Rad Laboratories, Richmond, CA). Endotoxin activity was contained in the first 1.0 ml of effluent after the void volume. This procedure was used for murine plasma and has not been applied to human plasma. The authors used spectrophotometric detection to determine the rate of increase in turbidity and reported an increase in sensitivity of the LAL assay of at least 10^4-fold. A plot of 1/t (where t was the time for a change in absorbance from 0.01 to 0.03 at 380 nm) vs. endotoxin concentration gave a straight line.

Variation in the sensitivity of LAL for endotoxin depending on the method of preparation and storage of lysate has been well documented.[103,106] Sullivan and Watson[107] studied some of the factors affecting this variation. The lysate prepared during summer months was more sensitive than that prepared during the winter. Sensitivity was improved and variability was decreased when lysate preparations were extracted with chloroform. Addition of divalent cations further increased sensitivity. A combination of chloroform extraction and addition of divalent cation (0.02 *M* $CaCl_2$) with 0.154 *M* NaCl increased sensitivity 100-fold. The effect of calcium was interesting, since it had been reported previously that disodium ethylenediaminetetraacetate (EDTA) did not inhibit the reaction,[99] although this finding was disputed by Reinhold and Fine.[105] The polyanionic compound, sodium polystyrolsulfonate, has been reported to inhibit the LAL reaction with endotoxin.[108]

Specificity

The LAL test has been shown to be highly specific for endotoxin. Rojas-Corona et al.[97] showed that several vasoactive compounds (Table 6) did not cause coagulation of LAL at concentrations of 10 to 100 times normal levels in serum. Substances that have been shown to lack endotoxin-like activity are listed in Table 6. Elin et al.[109] reported that thrombin showed endotoxin-like activity. Yin[100] disputed this finding and suggested that the earlier results were probably due to an impurity. Reinhold and Fine[105] also reported that thrombin did not cause gelation of LAL. Human leukocyte pyrogen (Table 6) is an interesting exception, since it does have endotoxin-like activity in rabbits.[109] In addition to substances in Table 6, broth cultures of several Gram-positive cocci and yeasts have been shown not to cause gelation of LAL.[105,108,110]

TABLE 6

Substances Lacking Endotoxin-like Activity in the LAL Test

Substance	Ref.
Serotonin; histamine; epinephrine; norepinephrine; bradykinin	97
Hemoglobin; thrombin;[a] calcium; streptolysin; streptodornase; streptokinase; *Clostridium tetani* toxin	105
Human leukocyte pyrogen; dextran	109

[a]Disputed result.

Endotoxin-like activity in the LAL test has been shown for peptidoglycan,[108] group A streptococcal exotoxins,[111] thromboplastin,[109] ribonuclease,[109] and polynucleotides.[109] In addition, trypsin has been shown to cause gelation by direct action on the coagulable protein component.[112] The activity of peptidoglycan and streptococcal exotoxins is several thousandfold less than endotoxin, and it is unlikely that any of these substances are a clinically significant cause of false-positive reactions.[108,111] Goldstein et al.[113] studied the specificity of the LAL test for neonates. They concluded that the development of intestinal flora did not lead to false-positive tests.

Clinical Studies

Clinical studies to evaluate the usefulness of the LAL test for endotoxemia have had mixed results. Levin et al.[114] reported results from a series of 98 patients in whom Gram-negative septicemia was suspected. Fifteen patients had positive blood cultures for Gram-negative bacteria, and the LAL test was positive in 10 of these patients. A site of infection with Gram-negative bacteria was found in 4 of the 5 patients who had a positive LAL test and negative blood cultures. The LAL test was positive in 2 additional patients with *Candida* fungemia, and each of these had a Gram-negative bacterial infection. Fifty-four patients with other illnesses, including 7 with fever of undetermined origin and 10 with bacteremia due to Gram-positive cocci, had negative tests. An additional 77 normal blood donors were also negative. Except for the 5 patients who had Gram-negative bacteremia with negative LAL tests, no patient that had a Gram-negative bacterial infection and negative LAL test was described. These results showed an apparent sensitivity for Gram-negative bacteremia of 75% and only one false poisitive in a patient with suspected Gram-negative bacteremia without demonstrated Gram-negative infection. Three of 4 patients with strongly positive tests, consistent with an endotoxin concentration of 5 ng/ml, died. These studies were extended to 281 patients with suspected Gram-negative bacteremia.[115] Thirty patients had Gram-negative bacteremia of whom 16 (53%) had a positive LAL test. Nineteen patients had a positive LAL test without Gram-negative bacteremia. These patients all had Gram-negative infections or died. Four additional patients with positive LAL tests had *Candida* fungemia. Two of these had Gram-negative bacterial infections, and one had infarction of the small bowel. The LAL test was negative in 14 patients with Gram-negative bacteremia and 25 patients with Gram-positive bacteremia. A significant relationship between a positive LAL test and mortality was found for patients with Gram-negative bacteremia ($p < 0.06$). There was also a significant relationship between a positive LAL and hypotension in all patients in the study ($p < 0.05$). Based on these data, Levin and coworkers[114,115] concluded that the LAL test was a clinically useful test for endotoxemia and that it had prognostic value. They emphasized[115] that sepsis and endotoxemia need not occur together. Oberle et al.[116] evaluated the LAL test for detection of Gram-negative bacterial sepsis in 18 severely malnourished children. Of 53 individual tests, 15 were positive.

Caridis et al.[117] presented data from 24 selected patients in whom endotoxemia was detected by the LAL test. Twelve patients had a septic cause, such as pneumonia or wound sepsis, and 12 had a nonseptic cause, such as heatstroke with hepatic failure, cirrhosis, or myeloblastic leukemia. They concluded that in patients with sepsis, endotoxemia was transient and disappeared shortly after the sepsis was eliminated. In patients without sepsis, endotoxemia occurred when the permeability of the intestinal wall was increased, such as by ishemic injury or peritonitis, or when the reticuloendothelial system was damaged, such as by cirrhosis, cytotoxic agents, or prolonged hypovolemia. Endotoxemia persisted in these patients, whose clinical course was characterized by progressive vascular collapse, shock lung syndrome, and gastrointestinal bleeding. These results indicate that the presence and persistence of endotoxemia may be clinically significant, whether or not Gram-negative bacterial infection is demonstrated. Levin[118] has suggested that endotoxemia may be responsible for the physiological changes that occur in sepsis, and that sepsis and bacteremia should be defined independently.

In contrast to these results, several groups have concluded that the LAL test for endotoxemia is of limited clinical usefulness. Martinez-G et al.[119] reported that of 38 patients with Gram-negative bacterial infections, 15 had positive blood cultures, and only 2 had positive LAL tests. One of the patients with a positive LAL test had bacteremia. In a study of 344 patients on whom blood cultures were requested, Stumacher et al.[120] obtained positive LAL tests in 28 of 65 patients with Gram-negative bacteremia and in 11 of 43 patients with localized Gram-negative bacterial infections. A positive test was obtained from 8 of 22 patients who had Gram-positive bacteremia without Gram-negative bacterial infection, as well as 21 patients in whom an infection was not found. No correlation between a positive endotoxin test and prognosis could be made. Similarly, Feldman et al.[121] and Elin et al.[106] concluded from their studies that the LAL assay for endotoxemia lacked clinical usefulness. This conclusion was based largely on low sensitivity of the test for detection of Gram-negative bacteremia and the need to perform multiple tests.[122] With two exceptions, Gram-negative bacterial infections were presented in all patients on whom a positive test was obtained. These exceptions occurred with ambulatory patients who were receiving radiation therapy.[106]

The significance of positive tests in the absence of infection by Gram-negative bacteria is uncertain. Evidence shows that the LAL test is specific for endotoxin except in circumstances that would be clinically unlikely. It has been suggested that these so called false-positive tests could be due to increased liberation of endogenous endotoxin or decreased clearance.[106,117] Failure to demonstrate endotoxin in the presence of Gram-negative bacterial infection may be explained on the basis of a lack of sensitivity, although in the studies where this problem was greatest, there were no obvious deficiencies in the procedures used. Better standardization of the LAL test will help resolve this question. There is clearly a controversy with regard to the prognostic value of detection of endotoxemia.

A much clearer picture has evolved for the usefulness of the LAL test for detection of Gram-negative bacterial meningitis.[124-126] Nachum et al.[125] obtained positive assays in the initial cerebrospinal fluid (CSF) specimen from 38 patients with Gram-negative bacterial meningitis. Gram stain was positive in only 25 cases; no false-positive tests occurred with CSF from 74 patients with Gram-positive bacterial meningitis. Subsequent studies have confirmed the specificity of this test. McCracken et al.[124] detected endotoxin in 71% of specimens obtained at the time of diagnosis. Several species of Gram-negative bacteria were isolated. A direct correlation between the presence, persistence, and concentration of endotoxin with prognosis was possible. In contrast, Berman et al.[123] have confirmed the sensitivity and specificity of the test, but found no correlation between initial CSF levels of endotoxin and the severity of disease at the time of diagnosis or to prognosis.

Jorgensen et al.[127] have used the LAL endotoxin assay to detect significant Gram-negative bacteriuria in children. They found that a positive test with urine diluted 1:100 or 1:1000 correlated with a bacterial count of $>10^5$ CFU/ml. From 209 urine specimens, they detected all 23 of 23 cases of significant bacteriuria due to Gram-negative bacteria.

Gas-liquid Chromatography

Gas-liquid chromatography (GLC) can be used for the analysis of complex mixtures of compounds that are sufficiently volatile to be transported through a separation column in the gas phase. Many compounds that are nonvolatile can be converted to volatile derivatives suitable for analysis. Mixtures of compounds produced by bacterial metabolism of various substrates or by chemical degradation of bacterial cells may thus be analyzed. Abel et al.[128] were among the first to investigate the feasibility of using GLC to identify microorganisms by analysis of differences in their chemical composition. Since that time, GLC has been developed into a useful technique for identification of bacteria. In particular, GLC is in widespread use in clinical laboratories for identification of anaerobic bacteria.[129] Several recent reviews of the application of GLC to detection and identification of microorganisms have been published.[3,130-132]

GLC can be a sensitive method for detection of trace amounts of material. Sullivan et al.[23] investigated whether GLC would allow more rapid detection of microbial growth in a conventional blood culture. Portions of the culture medium were acidified and analyzed for possible characteristic metabolic products. Electron capture, flame ionization, and thermal conductivity detectors were used. Unfortunately, analysis by GLC did not expedite detection of bacteremia.

The most exciting application of GLC in clinical microbiology is for detection and identification of bacteria directly from a clinical specimen. In 1970, Mitruka et al.[134] reported that analysis of serum extracts allowed the detection and identification of bacteremia in experimentally infected mice. Experiments using both facultatively anaerobic and anaerobic bacteria were reported. In each case, peaks that were characteristic of the infecting organism appeared in the chromatogram of serum obtained after infection. Most of these peaks were also detected in spent culture media from the same isolate. Inocula of 10^9 organisms per milliliter were used to infect the mice, and at the time of testing, the levels of bacteremia were between 5×10^3 and 4×10^5 organisms per milliliter. Such high levels of bacteremia are not common in the clinical situation; however, these experiments were later extended to detection of clinically significant bacterial infections. Sera from 84 patients with fever were obtained.[135] The infecting organisms and numbers of patients are shown in Table 7. Sera from 10 healthy individuals were used as controls. All sera were subjected to acid hydrolysis and derivatization before analysis. Chromatograms of

TABLE 7

Infections Identified by Gas-liquid Chromatography of Serum

Organism	Number of patients[a]	Clinical diagnosis
Streptococcus pneumoniae	21 (21)	Pneumonia
Klebsiella pneumoniae	6 (6)	Pneumonia
Acinetobacter calcoaceticus[b]	5 (3)	Pneumonia
Streptococcus spp.	26 (13)	Acute pharyngitis
Mycobacterium tuberculosis	10 (0)	Pulmonary tuberculosis
Escherichia coli	12 (5)	Pyelonephritis
Pseudomonas spp.	4 (4)	Bacteremia

[a]Number in parentheses is the number of patients with bacteremia.
[b]*Herellea vaginicola* in the original reference.

Data derived from Mitruka, B. M., Kundargi, R. S., and Jonas, A. M., *Med. Res. Eng.*, 11, 7, 1972. With permission.

sera from each type of infection were reported to have between two and five characteristic peaks. Most of these peaks were also detected in spent culture media from the same organism; however, a few additional peaks were detected in sera, but not in culture media or control sera. It was postulated that these peaks may be due to substances produced as a result of host response to infectious agents. These experiments were extended to a study of 80 coded serum samples. These samples consisted of both normal serum and serum from patients with pneumococcal pneumonia. It was reported[135] that pneumococcal pneumonia was detected with greater than 90% accuracy. Additional details of this experiment were not given. Successful application of similar methods to detection of infections due to other organisms has been reported now by Mitruka.[3,136] Some of the peaks that appear in chromatograms of serum from patients with pneumococcal infections have been identified and are presumably components of the capsular polysaccharide antigens from these organisms.[3,136] As a result, Mitruka has not only been able to detect pneumococcal infection by analysis of serum specimens, but he can possibly tell which serotype.[136] The procedures have also been used to identify the infectious agent in patients that are not bacteremic.[3] Practical application of this exciting new procedure will require confirmation by other investigators.

Davis and McPherson[137] have also studied detection of septicemia by GLC. In their study, serum from patients with bacteremia was found to be indistinguishable from serum of normal controls. The procedure used was similar, but not identical to the procedure used by Mitruka.[3] A significant difference may have been the use of flame-ionization detection rather than the more sensitive electron-capture detection.[137] Davis and McPherson,[137] however, were able to detect *Candida* septicemia in all of 13 cases they studied. Total time for detection and identification was reported to be less than 3 hr.

GLC has been used to detect bacterial and fungal infections in clinical specimens other than serum. Brooks et al.[138] have shown that metabolic products from *Proteus mirabilis* could be detected in urine from two patients with bacteriuria due to this organism. These compounds disappeared following successful antibiotic treatment. One of the compounds – N-nitrosodimethylamine – was of particular interest since it is a potent carcinogen, and the exposure of one patient was calculated to be 0.8 g over a period of 11 years.[138] In another study, Brooks et al.[139] used GLC of synovial fluid extracts to differentiate between traumatic arthritis and septic arthritis due to *Neisseria gonorrhoeae, S. aureus,* and *S. pyogenes.* Differentiation between the various organisms was possible, but the number of cases studied was limited. Contronі et al.[140] have reported that detection of lactic acid in CSF by GLC is a sensitive and specific test for bacterial meningitis. Amundson et al.[141] have shown by GLC analysis that cryptococcal meningitis could be identified by a reduction in the concentration of two compounds normally present in CSF.

Several groups[142–144] have used GLC for direct analysis of short-chain fatty acids (SCFA) in purulent material. Infections involving only aerobic bacteria contained acetic acid; whereas, when anaerobic bacteria were involved, a mixture of SCFA was detected. The presence of isobutyric acid, butyric acid, or succinic acid was sufficient to indicate the presence of anaerobic bacteria. False-positive and false-negative results were rare.

Counterimmunoelectrophoresis

Counterimmunoelectrophoresis (CIE) is a recently popularized immunologic technique for detection of soluble antigens in various fluids. The presence of detectable bacterial antigen has been shown to correlate well with the presence of disease and other laboratory findings suggestive of infection. Wadstrom and Nord[145] and Rytel[146] have recently reviewed the history and application of CIE to diagnosis of infectious diseases, and an extensive review of immunoelectrophoretic methods has been made by Verbruggen.[147] CIE has been applied most widely to the detection of meningitis; however, the scope of this section will be to survey the application of CIE to detection of antigenemia and bacteremia.

Theory

Most bacterial polysaccharide capsular antigens are negatively charged in slightly alkaline media (pH 8.2 to 8.6). Under the same conditions, antibodies are neutral or only slightly charged. In CIE, two adjacent wells are placed in an agarose-coated slide. One well is filled with the fluid to be tested for antigen, and the other is filled with antiserum. An electric field is applied across the slide such that the antigen, which migrates toward the positive electrode (anode), will enter a zone of reaction between the two wells. The antibody, which is carried in the opposite direction toward the negative electrode by the normal flow of buffer (electroendosmosis), also enters the zone of reaction. With specific antigen and antibody, the resulting antigen-antibody complex forms a precipitin line between the two wells. By varying the antibody, the presence and identity of soluble bacterial antigens can be determined. Conversely, the presence of specific antibody can be determined by using a known preparation of antigen. Buffers that consist principally of barbital are most frequently used; however, capsular antigens from pneumococcal types VII and XIV are not detectable with barbital buffer alone.[148,149] Addition of sulfonated phenylboronic acid or *m*-carboxyphenylboronic acid to the barbital buffer allows detection of these two pneumococcal serotypes.[148,150]

Haemophilus influenzae

Detection of *H. influenzae* antigen in serum and CSF has been of particular interest for diagnosis of meningitis;[151–157] however, antigenemia has been detected in cases of epiglottitis and pericarditis as well.[158] In most studies in which antigenemia was detected, data regarding the presence of bacteremia were not reported. Smith and Ingram[158] found that in 8 patients with epiglottitis and bacteremia, antigenemia was detected in 5 patients. Feigin et al.[153] reported that of 50 patients with *Haemophilus* meningitis, 41 (82%) had positive blood cultures and 25 (50%) had antigenemia. There was a significant correlation between prolonged antigenemia and prolonged fever and sequelae of meningitis ($p < 0.01$). Shackelford et al.[157] reported that of 18 patients with *Haemophilus* meningitis, bacteremia was present in 15 patients and antigenemia was detected in 8 patients. For patients without complications or sequelae, the mean serum antigen level was 70 ng/ml, whereas patients with significant sequelae had a mean concentration of 850 ng/ml. These differences were significant ($p < 0.025$). Ingram et al.[156] found that prolonged antigenemia correlated with more severe symptoms, prolonged fever, and lower antibody responses, while elevated levels of antigenemia alone were not associated with these findings.

Streptococcus pneumoniae

Antigenemia has been detected in pneumococcal meningitis[154,157,159] and pneumococcal pneumonia.[149,160–167] Antigenemia correlates well with bacteremia in pneumococcal pneumonia (Table 8), but sensitivity is low. Antigenuria has been detected more frequently than antigenemia in bacteremic pneumonia. It has also been detected more frequently in nonbacteremic pneumonia (Table 8). Two groups have reported that very high levels of antigenemia were detected in a small number of patients who died.[149,169] Coonrod and Rytel[161] have reported that the presence or absence of antigenemia

TABLE 8

Correlation of Pneumococcal Polysaccharide Detection in Serum and Urine with Bacteremia

Clinical diagnosis	Number of patients	Number of patients positive[a,b] Blood culture	Serum CIE	Urine CIE	Comment	Year and ref.
Pneumonia	17	17	3	NT		1971[164]
Pneumonia	20	20	12	NT	Monovalent antisera used; with polyvalent antiserum only 5 of 12 positive sera were detected	1972[149]
Pneumonia	26	0	0	NT		
Meningitis	6	4	3 (4)	3 (3)	Antigen in CSF in 5 of 6 cases; three patients positive by all three methods	1973[159]
Pneumonia	11	11	5	7	Urine concentrated 20-fold	1973[161]
Pneumonia	19	0	1	7		
Pneumonia	10	10	8	4 (7)		1974[160]
Pneumonia	12	0	4	1 (4)		
Bronchitis	5	0	2	NT		
Pneumonia	10	10	7 (NS)	8 (8)	Polysaccharide antigen in urine was of lower molecular weight than antigen in serum	1974[162]
Pneumonia	98	20	28	42 (78)		1975[167]
Pneumonia	245	Rare	0	NT	Study involved outpatients only	1975[168]
Pneumonia	92	NR	32	NT	All patients had pneumococci isolated from sputum	1976[166]
Chronic chest disease	83	0 (?)	1	NT		
Post-operative	51	0 (?)	0	NT		
Pneumonia	26	26	17	NT		1976[163]
Pneumonia	20	0	2	NT		
Pneumonia	8	8	3 (5)	2 (4)		1976[165]
Pneumonia	11	0	0 (10)	0 (10)		

[a]Abbreviations: NT, not tested; NR, not reported.
[b]Number in parentheses is number of cases tested if different from total.

correlated with pleural effusions, azotemia, and hyperbilirubinemia ($p < 0.05$). Spencer and Savage[166] reported that although the level of antigenemia did not correlate with poor prognosis, the presence of antigenemia did. In their study, 50% of patients with antigenemia died compared to 10% without antigenemia. Coonrod and Drennan[163] reported that antigenemia was associated with delayed antibody response and a severe and protracted illness. In contrast, Tugwell and Greenwood[167] reported that antigenemia correlated with jaundice, diarrhea, and pyrexia, but not with any other clinical feature or death.

Neisseria meningitidis

N. meningitidis antigens have been detected in the serum of patients with meningococcal meningitis either with or without bacteremia,[170-173] and antigenemia has generally been found to be a poor prognostic sign. Whittle et al.[173] detected antigen in serum from 27 of 200 patients with meningitis. Antigenemia was present in 7 of 33 patients with bacteremia and in 16 of 137 patients without bacteremia. The difference in detection rates in bacteremic and nonbacteremic patients was not statistically significant. However, those patients with antigenemia had a worse prognosis with more frequent complications. Interestingly, antigen was detected in CSF in 129 of these patients. There was no correlation between duration of disease and antigen level in CSF, but there was a correlation between antigen level and severity of disease ($p < 0.001$). Edwards[170] also found that antigenemia was a poor prognostic sign. Of 23 patients with meningitis, all of 5 patients with antigenemia died. Hoffman and Edwards[172] showed that there was an inverse relationship between level of antigenemia and the level of fibrinogen and the number of platelets and white blood cells. The appearance of antibodies in patients with meningococcal infection without antigenemia occurred significantly sooner than in patients with antigenemia ($p < 0.01$).

One problem in diagnosis of meningococcal infections with CIE has been the difficulty in obtaining high-titered antisera. This has been of particular problem in diagnosis of infections due to group B meningococci.[174]

Escherichia coli

McCracken et al.[175] reported that antiserum prepared against *N. meningitidis* group B could be used in CIE to detect capsular polysaccharide antigen from *E. coli* K1 in serum and CSF of infants with *E. coli* meningitis. CIE detected K1 antigen in CSF of 29 of 41 patients with meningitis due to *E. coli* K1. A direct correlation between severity of disease and presence of K1 antigen was found. Serum from 36 of these patients was available for study. Capsular antigen was detectable in 15 (42%) of these sera. Death occurred in 73% of those patients with antigenemia compared to 14% of those patients without antigenemia.

Staphylococcal Teichoic Acid

Jackson et al.[176] have presented two cases in which staphylococcal teichoic acid was detected by CIE. One case had staphylococcal pericarditis, and antigen was demonstrated in the pericardial fluid. The other case had staphylococcal meningitis with antigen in the CSF. Lampe et al.[177] detected staphylococcal antigen in pleural fluid from which *Staphylococcus aureus* was isolated.

Group B Streptococci

Hill et al.[178] reported detection of group B streptococcal antigen by CIE in CSF and peritoneal fluid from infected patients. They were unable to detect antigen in the serum from a third patient with group B streptococcal sepsis. Hill[179] was also unable to detect group B streptococcal antigen in blood that had been inoculated with group B streptococci and incubated for 4 to 24 hr. The reason for this inability to detect group B antigen in serum was unexplained.

Pseudomonas aeruginosa

Bartram et al.[160] reported that of 12 patients with severe infections due to *P. aeruginosa*, antigenemia was detected in 11 cases by CIE. Only 5 of the 12 patients had positive blood cultures. They found that weak cross-reactions could be demonstrated with extracts of *Proteus* spp., *Klebsiella pneumoniae, E. coli,* and *Enterobacter aerogenes.* Serum from 16 patients with serious infections, 14 of whom had bacteremia due to *Enterobacter* spp., group C streptococci, staphylococci, *H. influenzae, E. coli, Serratia* spp., *Bacteroides fragilis,* or *Streptococcus pneumoniae,* were tested. Despite the evidence of crossreactions with bacterial extracts, no crossreactions were found using serum specimens. As an interesting note, they were also able to detect antigenemia in 4 patients with infections due to Enterobacteriaceae when antiserum against the infecting organism was used. Adams et al.[180] have also reported detection of antigenemia in patients with systemic infections due to *P. aeruginosa.* In this study, however, serum from 2 of 28 patients without *Pseudomonas* infection also gave positive reactions. The identities of the organisms isolated from these patients were not reported.

Klebsiella pneumoniae K2

Simpson and Speller[181] detected the capsular K2 antigen in serum from 3 of 3 patients with bacteremia due to *K. pneumoniae* K2. Although antigenuria was present in 21 of 21 patients with transient or persistent urinary tract colonization, antigenemia could not be detected. These results helped provide rapid diagnosis of bacteremia during an outbreak of *K. pneumoniae* K2 infections in a surgical ward.

Specificity

Similarities in the antigenic structure of capsular antigens from different bacteria are well known. Crossreactions between various antisera and bacterial antigens produced in vitro have been demonstrated by CIE;[160,182] although, crossreactions using clinical specimens have been only rarely reported. Shackelford et al.[157] reported a positive reaction of group C meningococcal antiserum with serum from a patient with group B meningococcal bacteremia. In addition, they reported four possible false-positive reactions with *N. meningitidis* group D antiserum. Lampe et al.[177] reported one case in which a cross-reaction between *Escherichia coli* antigen and *H. influenzae* type b antiserum occurred. This reaction was observed with pleural fluid, which presumably contained a high concentration of antigen. Ribner et al.[183] reported one case in which a concentrated CSF specimen that contained *Staphylococcus aureus* antigen cross-reacted with *H. influenzae* type b antiserum. More recently, Colding and Lind[184] detected a crossreaction between *E. coli* antigen and *H. influenzae* type b antiserum in a CSF specimen. Ingram et al.[155] have shown that cross reactions of broth supernatants from *E. coli* and *Streptococcus pneumoniae* with *Haemphilus* type b antisera may be eliminated by absorption. There was no loss in sensitivity for detection of *Haemophilus* type b antigen. Not all cross-reactions, however, are detrimental. McCracken et al.[175] showed that antigen in CSF and serum from meningitis cases due to *E. coli* K1 could be detected with meningococcal group B antiserum. This crossreaction was used to provide useful diagnostic information and did not create a diagnostic problem.

REFERENCES

1. **Prier, J. E., Bartola, J. T., and Friedman, H.,** *Modern Methods in Medical Microbiology,* University Park Press, Baltimore, MD, 1976.
2. **Heden, C.-G. and Illeni, T.,** *New Approaches to the Identification of Microorganisms,* John Wiley & Sons, New York, 1975.
3. **Mitruka, B. M.,** *Methods of Detection and Identification of Bacteria,* CRC Press, Cleveland, 1976.
4. **Hirsch, J. G.,** Comparative bactericidal activities of blood serum and plasma serum, *J. Exp. Med.,* 112, 15, 1960.
5. **Washington, J. A., II,** Blood cultures, principles, and techniques, *Mayo Clin. Proc.,* 50, 91, 1975.
6. **Traub, W. H. and Kleber, I.,** Inactivation of classical and alternative pathway-activated bactericidal activity of human serum by sodium polyanetholsulfonate, *J. Clin. Microbiol.,* 5, 278, 1977.
7. **Sullivan, N. M., Sutter, V. L., and Finegold, S. M.,** Practical aerobic membrane filtration blood culture technique: clinical blood culture trial, *J. Clin. Microbiol.,* 1, 37, 1975.
8. **Kagan, R. L., Schuette, W. H., Zierdt, C. H., and MacLowry, J. D.,** Rapid automated diagnosis of bacteremia by impedance detection, *J. Clin. Microbiol.,* 5, 51, 1977.
9. **Pickett, M. J. and Nelson, E. L.,** Observations on the problem of Brucella blood cultures, *J. Bacteriol.,* 61, 229, 1951.
10. **Braun, W. and Kelsh, J.,** Improved method for cultivation of *Brucella* from the blood, *Proc. Soc. Exp. Biol. Med.,* 85, 154, 1954.
11. **Dorn, G. L., Haynes, J. R., and Burson, G. G.,** Blood culture technique based on centrifugation: developmental phase, *J. Clin. Microbiol.,* 3, 251, 1976.
12. **Dorn, G. L., Burson, G. G., and Haynes, J. R.,** Blood culture technique based on centrifugation: clinical evaluation, *J. Clin. Microbiol.,* 3, 258, 1976.
13. **Tidwell, W. L. and Gee, L. L.,** Use of membrane filter in blood cultures, *Proc. Soc. Exp. Biol. Med.,* 88, 561, 1955.
14. **Vacek, V. and Svejcar, J.,** Prispevek k diagnostice septickych onemocneni. I. Hemokultivace metodou membranove filtrace krve [Diagnosis of septic conditions. I. Hemoculture method with membrane filtration], *Cas. Lek. Cesk.,* 97, 1281, 1958.
15. **Dodin, A., Brygoo, E. R., and Sureau, P.,** Recherche et numeration des enterobacteries par hemoculture sur membranes filtrantes, *Ann. Inst. Pasteur* (Paris), 96, 489, 1959.
16. **Randriambololona, R. and Dodin, A.,** Etude de l'evolution de la bacteriemie dans onze cas de salmonellose traites, *Ann. Inst. Pasteur* (Paris), 99, 278, 1960.
17. **Kozub, W. R., Kirkham, W. R., Chatman, C. E., and Pribor, H. C.,** A practical blood culturing method employing dilution and filtration, *Am. J. Clin. Pathol.,* 52, 105, 1969.
18. **Finegold, S. M., White, M. L., Ziment, I., and Winn, W. R.,** Rapid diagnosis of bacteremia, *Appl. Microbiol.,* 18, 458, 1969.
19. **Winn, W. R., White, M. L., Carter, W. T., Miller, A. B., and Finegold, S. M.,** Rapid diagnosis of bacteremia with quantitative differential-membrane filtration culture, *JAMA,* 197, 111, 1966.
20. **Stanaszek, P. M.,** Rapid bacteremia diagnosis using field monitor membrane filtration, *Am. J. Med. Technol.,* 37, 97, 1971.
21. **Farmer, S. G. and Komorowski, R. A.,** Evaluation of the Sterifil lysis-filtration blood culture system, *Appl. Microbiol.,* 23, 500, 1972.
22. **Finegold, S. M., Sutter, V. L., and Carter, W. T.,** An improved membrane filtration system for blood cultures, *Bacteriol. Proc.,* 106, 1969.
23. **Sullivan, N. M., Sutter, V. L., Carter, W. T., Attebery, H. R., and Finegold, S. M.,** Bacteremia after genitourinary tract manipulation: bacteriological aspects and evaluation of various blood culture systems, *Appl. Microbiol.,* 23, 1101, 1972.

24. **Sullivan, N. M., Sutter, V. L., and Finegold, S. M.,** Practical aerobic membrane filtration blood culture technique: development of procedure, *J. Clin. Microbiol.,* 1, 30, 1975.
25. **Zierdt, C. H., Kagan, R. L., and MacLowry, J. D.,** Development of a lysis-filtration blood culture technique, *J. Clin. Microbiol.,* 5, 46, 1977.
26. **Rose, R. E. and Bradley, W. J.,** Using the membrane filter in clinical microbiology, *Med. Lab,* 3, 22, 1969.
27. **Schrot, J. R., Hess, W. C., and Levin, G. V.,** Method for radiorespirometric detection of bacteria in pure culture and in blood, *Appl. Microbiol.,* 26, 867, 1973.
28. **Harkness, J. L., Hall, M., Ilstrup, D. M., and Washington, J. A., II,** Effects of atmosphere of incubation and of routine subcultures on detection of bacteremia in vacuum blood culture bottles, *J. Clin. Microbiol.,* 2, 296, 1975.
29. **Komorowski, R. A. and Farmer, S. G.,** Rapid detection of candidemia, *Am. J. Clin. Pathol.,* 59, 56, 1973.
30. **Todd, J. K. and Roe, M. H.,** Rapid detection of bacteremia by an early subculture technic, *Am. J. Clin. Pathol.,* 64, 694, 1975.
31. **Sladek, K. J., Suslavich, R. V., Sohn, B. I., and Dawson, F. W.,** Optimum membrane structures for growth of coliform and fecal coliform organisms, *Appl. Microbiol.,* 30, 685, 1975.
32. **Levin, G. V., Harrison, V. R., and Hess, W. C.,** Preliminary report on a one-hour presumptive test for coliform organisms, *J. Am. Water Works Assoc.,* 48, 75, 1956.
33. **Levin, G. V., Harrison, V. R., Hess, W. C., and Gurney, H. C.,** A radioisotope technique for the rapid detection of coliform organisms, *Am. J. Public Health,* 46, 1405, 1956.
34. **Buddemeyer, E. U.,** Liquid scintillation vial for cumulative and continuous radiometric measurement of in vitro metabolism, *Appl. Microbiol.,* 28, 177, 1974.
35. **Buddemeyer, E., Hutchinson, R., and Cooper, M.,** Automatic quantitative radiometric assay of bacterial metabolism, *Clin. Chem.,* 22, 1459, 1976.
36. **DeLand, F. H. and Wagner, H. N., Jr.,** Early detection of bacterial growth, with carbon-14-labeled glucose, *Radiology,* 92, 154, 1969.
37. **DeLand, F. and Wagner, H. N., Jr.,** Automated radiometric detection of bacterial growth in blood cultures, *J. Lab. Clin. Med.,* 75, 529, 1970.
38. **Washington, J. A., II and Yu, P. K. W.,** Radiometric method for detection of bacteremia, *Appl. Microbiol.,* 22, 100, 1971.
39. **Waters, J. R.,** Sensitivity of the $^{14}CO_2$ radiometric method for bacterial detection, *Appl. Microbiol.,* 23, 198, 1972.
40. **Bannatyne, R. M. and Harnett, N.,** Radiometric detection of bacteremia in neonates, *Appl. Microbiol.,* 27, 1067, 1974.
41. **Brooks, K. and Sodeman, T.,** Rapid detection of bacteremia by a radiometric system, *Am. J. Clin. Pathol.,* 61, 859, 1974.
42. **DeBlanc, H. J., Jr., DeLand, F., and Wagner, H. N., Jr.,** Automated radiometric detection of bacteria in 2,967 blood cultures, *Appl. Microbiol.,* 22, 846, 1971.
43. **Morello, J. A.,** Automated radiometric detection of bacteremia, in *Automation in Microbiology and Immunology,* Heden, C.-G. and Illeni, T., Eds., John Wiley & Sons, New York, 1975.
44. **Randall, E. L.,** Long-term evaluation of a system for radiometric detection of bacteremia, in *Microbiology – 1975,* Schlessinger, D., Ed., American Society for Microbiology, Washington, D.C., 1975.
45. **Smith, A. G. and Little, R. R.,** Detection of bacteremia by an automated radiometric method and a tubed broth method, *Ann. Clin. Lab. Sci.,* 4, 448, 1974.
46. **Smith, A. G. and Little, R. R.,** The BACTEC 225: a radiometric technique for the detection of bacteremia, *Lab. Med.,* 6, 18, 1975.
47. **Thiemke, W. A. and Wicher, K.,** Laboratory experience with a radiometric method for detecting bacteremia, *J. Clin. Microbiol.,* 1, 302, 1975.
48. **Randall, E. L.,** Radiometric techniques in microbiology, in *Modern Methods in Medical Microbiology,* Prier, J. E., Bartola, J. T., and Friedman, H., Eds., University Park Press, Baltimore, MD, 1976.
49. **Zwarun, A. A.,** Agitation and the radiometric detection of clinically isolated aerobic bacteria, *Abstr. Ann. Meeting ASM,* M274, 1974.
50. BACTEC Application Note, JLI-623, January 1975, Johnston Laboratories, Cockeysville, MD.
51. BACTEC Application Note, JLI-620A, August 1976, Johnston Laboratories, Cockeysville, MD.
52. **Ellner, P. D., Kiehn, T. E., Beebe, J. L., and McCarthy, L. R.,** Critical analysis of hypertonic medium and agitation in detection of bacteremia, *J. Clin. Microbiol.,* 4, 216, 1976.
53. **Waters, J. R. and Zwarun, A. A.,** Results of an automated radiometric sterility test as applied to clinical blood cultures, *Dev. Ind. Microbiol.,* 14, 80, 1973.
54. **Caslow, M., Ellner, P. D., and Kiehn, T. E.,** Comparison of the BACTEC system with blind subculture for the detection of bacteremia, *Appl. Microbiol.,* 28, 435, 1974.
55. **Previte, J. J., Rowley, D. B., and Wells, R.,** Improvements in a non-proprietary radiometric medium to allow detection of some *Pseudomonas* species and *Alcaligenes faecalis, Appl. Microbiol.,* 30, 339, 1975.
56. **Larson, S. M., Charache, P., Chen, M., and Wagner, H. N., Jr.,** Automated detection of *Haemophilus influenzae, Appl. Microbiol.,* 25, 1011, 1973.

57. **Rosner, R.**, Comparisons of isotonic and radiometric-hypertonic cultures for the recovery of organisms from cerebrospinal, pleural and synovial fluids, *Am. J. Clin. Pathol.*, 63, 149, 1975.
58. **Zwarun, A. A.**, Effect of osmotic stabilizers on $^{14}CO_2$ production by bacteria and blood, *Appl. Microbiol.*, 25, 589, 1973.
59. **Coleman, R. M., Laslie, W. W., and Lambe, D. W., Jr.**, Clinical comparison of aerobic hypertonic, and anaerobic culture media for the radiometric detection of bacteremia, *J. Clin. Microbiol.*, 3, 281, 1976.
60. **Hull, K. H., Silvanic, J. M., and Wolsh, L. J.**, Evaluation of BACTEC hypertonic (8A) medium, *Abstr. Ann. Meeting ASM*, C77, 1976.
61. **Kulas, C. M., Short, H. B., Speck, E. L., Betts, R. F., and Robertson, R. G.**, An evaluation of the efficacy of ^{14}C-labeled osmotically stabilized blood culture media in an automated blood culture system, *Abstr. Ann. Meeting ASM*, M277, 1974.
62. **Rosner, R.**, Growth patterns of a wide spectrum of organisms encountered in clinical blood cultures using both hypertonic and isotonic media, *Am. J. Clin. Pathol.*, 65, 706, 1976.
63. **Washington, J. A., II, Hall, M. M., and Warren, E.**, Evaluation of blood culture media supplemented with sucrose or with cysteine, *J. Clin. Microbiol.*, 1, 79, 1975.
64. **Rosner, R.**, Comparison of macroscopic, microscopic, and radiometric examinations of clinical blood cultures in hypertonic media, *Appl. Microbiol.*, 28, 644, 1974.
65. BACTEC Application Note, JLI-456H, October 1976, Johnston Laboratories, Cockeysville, MD.
66. **French, J. E., Crum, L. P., Groschel, D., and Hopfer, R. L.**, Experience with the BACTEC 225 blood culture system in a cancer hospital, *Abstr. Ann. Meeting ASM*, C78, 1976.
67. BACTEC Application Note, JLI-619A, August 1976, Johnston Laboratories, Cockeysville, MD.
68. **Wood, N. G.**, Comparative study of two systems for detecting bacteraemia and septicaemia, *J. Clin. Pathol.*, 29, 530, 1976.
69. **Renner, E. D., Gatherdige, L. A., and Washington, J. A., II**, Evaluation of radiometric system for detecting bacteremia, *Appl. Microbiol.*, 26, 368, 1973.
70. **Blazevic, D. J., Stemper, J. E., and Matsen, J. M.**, Comparison of macroscopic examination, routine Gram stains, and routine subcultures in the initial detection of positive blood cultures, *Appl. Microbiol.*, 27, 537, 1974.
71. **Hall, M., Warren, E., and Washington, J. A., II**, Detection of bacteremia with liquid media containing sodium polyanetholsulfonate, *Appl. Microbiol.*, 27, 187, 1974.
72. **Hall, M. M., Ilstrup, D. M., and Washington, J. A., II**, Effect of volume of blood cultured on detection of bacteremia, *J. Clin. Microbiol.*, 3, 643, 1976.
73. **Strauss, R. R., Throm, R., and Friedman, H.**, Radiometric detection of bacteremia: requirement for terminal subcultures, *J. Clin. Microbiol.*, 5, 145, 1977.
74. **Stewart, G. N.**, The changes produced by the growth of bacteria in the molecular concentration and electrical conductivity of culture media, *J. Exp. Med.*, 4, 235, 1899.
75. **Hadley, W. K. and Senyk, G.**, Early detection of microbial metabolism and growth by measurement of electrical impedance, in *Microbiology – 1975*, Schlessinger, D., Ed., American Society for Microbiology, Washington, D.C., 1975.
76. **Cady, P.**, Rapid automated bacterial identification by impedance measurement, in *New Approaches to the Identification of Microorganisms*, Heden, C.-G. and Illeni, T., Eds., John Wiley & Sons, New York, 1975.
77. **Stoner, G. E., Wilkins, J. R., and Lemeland, J. F.**, Automated microbial detection and quantification, *J. Electrochem. Soc.*, 122, 109C (abstr. 382), 1975.
78. **Ur, A. and Brown, D. F. J.**, Impedance monitoring of bacterial activity, *J. Med. Microbiol.*, 8, 19, 1975.
79. **Ur, A. and Brown, D. F. J.**, Monitoring of bacterial activity by impedance measurements, in *New Approaches to the Identification of Microorganisms*, Heden, C.-G. and Illeni, T., Eds., John Wiley & Sons, New York, 1975.
80. **Ur, A. and Brown, D. F. J.**, Rapid detection of bacterial activity using impedance measurements, *BioMed. Eng.*, 9, 18, 1972.
81. **Cady, P. and Dufour, S. W.**, Automated detection of microorganism metabolism and growth by impedance measurements, *Abstr. Ann. Meeting ASM*, E43, 1974.
82. **Hadley, W. K., Senyk, G., and Michaels, R.**, Early detection of bacterial growth by measurements of electrical impedance in blood cultures, *Abstr. Ann. Meeting ASM*, M281, 1974.
83. **Hadley, W. K. and Kazinka, W.**, Comparison of impedance measurements and standard laboratory procedures for detection of microorganisms in blood cultures, *Abstr. Ann. Meeting ASM*, C69, 1976.
84. **Khan, W., Friedman, G., Rodriguez, W., Controni, G., and Ross, S.**, Rapid detection of bacteria in blood and spinal fluids in children by electrical impedance method, *Abstr. Ann. Meeting ASM*, C70, 1976.
85. **Forrest, W. W.**, Microcalorimetry, in *Methods in Microbiology*, Vol. 6B, Norris, J. R. and Ribbons, D. W., Eds., Academic Press, New York, 1972, 285.
86. **Boling, E. A., Blanchard, G. C., and Russell, W. J.**, Bacterial identification by microcalorimetry, *Nature*, 241, 472, 1973.
87. **Russell, W. J., Zettler, J. F., Blanchard, G. C., and Boling, E. A.**, Bacterial identification by microcalorimetry, in *New Approaches to the Identification of Microorganisms*, Heden, C.-G. and Illeni, T., Eds., John Wiley & Sons, New York, 1975.

88. **Russell, W. J., Farling, S. R., Blanchard, G. C., and Boling, E. A.,** Interim review of microbial identification by microcalorimetry, in *Microbiology–1975,* Schlessinger, D., Ed., American Society for Microbiology, Washington, D.C., 1975.
89. **Ripa, K. T., Mardh, P.-D., Hovelius, B., and Ljungholm, K.,** Microcalorimetry as a tool for evaluation of blood culture media, *J. Clin. Microbiol.,* 5, 393, 1977.
90. **McElroy, W. D., Seliger, H. H., and White, E. H.,** Mechanism of bioluminescence, chemiluminescence and enzyme function in the oxidation of firefly luciferin, *Photochem. Photobiol.,* 10, 153, 1969.
91. **Chappelle, E. W. and Levin, G. V.,** Use of the firefly bioluminescent reaction for rapid detection and counting of bacteria, *Biochem. Med.,* 2, 41, 1968.
92. **Alexander, D. N., Ederer, G. M., and Matsen, J. M.,** Evaluation of an adenosine 5′-triphosphate assay as a screening method to detect significant bacteriuria, *J. Clin. Microbiol.,* 3, 42, 1976.
93. **Conn, R. B., Charache, P., and Chappelle, E. W.,** Limits of applicability of the firefly luminescence ATP assay for the detection of bacteria in clinical specimens, *Am. J. Clin. Pathol.,* 63, 493, 1975.
94. **Thore, A., Ansehn, Lundin A., and Bergman, S.,** Detection of bacteriuria by luciferase assay of adenosine triphosphate, *J. Clin. Microbiol.,* 1, 1, 1975.
95. **Schrock, C. G., Barza, M. J., Deming, J. W., Picciolo, G. L., Chappelle, E. W., and Weinstein, L.,** Rapid detection of bacterial growth in blood cultures by means of adenosine triphosphate determination, *Abstr. Ann. Meeting ASM,* C71, 1976.
96. **Levin, J. and Bang, F. B.,** Clottable protein in Limulus: its localization and kinetics of its coagulation by endotoxin, *Thromb. Diath. Haemorrh.,* 19, 186. 1968.
97. **Rojas-Corona, R. R., Skarnes, R., Tamakuma, S., and Fine, J.,** The *Limulus* coagulation test for endotoxin. A comparison with other assay methods, *Proc. Soc. Exp. Biol. Med.,* 132, 599, 1969.
98. **Cooper, J. F., Levin, J., and Wagner, H. N., Jr.,** Quantitative comparison of in vitro and in vivo methods for the detection of endotoxin, *J. Lab. Clin. Med.,* 78, 138, 1971.
99. **Young, N. S., Levin, J., and Prendergast, R. A.,** An invertebrate coagulation system activated by endotoxin: evidence for enzymatic mediation, *J. Clin. Invest.,* 51, 1790, 1972.
100. **Yin, E. T.,** Endotoxin, thrombin, and the *Limulus* amebocyte lysate test, *J. Lab. Clin. Med.,* 86, 430, 1975.
101. **Selzer, G. B.,** Preparations of a purified lipopolysaccharide for pyrogen testing, *Bull. Parenter. Drug Assoc.,* 24, 153, 1970.
102. **Rudbach, J. A., Akiya, F. I., Elin, R. J., Hochstein, H. D., Luoma, M. K., Milner, E. C. B., Milner, K. C., and Thomas, K. R.,** Preparation and properties of a national reference endotoxin, *J. Clin. Microbiol.,* 3, 21, 1976.
103. **Levin, J., Tomasulo, P. A., and Oser, R. S.,** Detection of endotoxin in human blood and demonstration of an inhibitor, *J. Lab. Clin. Med.,* 75, 903, 1970.
104. **Hollander, V. P. and Harding, W. C.,** A sensitive spectrophotometric method for measurement of plasma endotoxin, *Biochem. Med.,* 15, 28, 1976.
105. **Reinhold, R. B. and Fine, J.,** A technique for quantitative measurement of endotoxin in human plasma, *Proc. Soc. Exp. Biol. Med.,* 137, 334, 1971.
106. **Elin, R. J., Robinson, R. A., Levine, A. S., and Wolff, S. M.,** Lack of clinical usefulness of the limulus test in the diagnosis of endotoxemia, *N. Engl. J. Med.,* 293, 521, 1975.
107. **Sullivan, J. D., Jr. and Watson, S. W.,** Factors affecting the sensitivity of *Limulus* lysate, *Appl. Microbiol.,* 28, 1023, 1974.
108. **Wildfeuer, A., Heymer, B., Schleifer, K. H., and Haferkamp, O.,** Investigations on the specificity of the *Limulus* test for the detection of endotoxin, *Appl. Microbiol.,* 28, 867, 1974.
109. **Elin, R. J. and Wolff, S. M.,** Nonspecificity of the limulus amebocyte lysate test: positive reactions with polynucleotides and proteins, *J. Infect. Dis.,* 128, 349, 1973.
110. **Jorgensen, J. H. and Smith, R. F.,** Preparation, sensitivity and specificity of *Limulus* lysate for endotoxin assay, *Appl. Microbiol.,* 26, 43, 1973.
111. **Brunson, K. W. and Watson, D. W.,** Limulus amebocyte lysate reaction with streptococcal pyrogenic exotoxin, *Infect. Immun.,* 14, 1256, 1976.
112. **Solum, N. O.,** The coagulogen of *Limulus polyphemus* hemocytes. A comparison of the clotted and non-clotted forms of the molecule, *Thromb. Res.,* 2, 55, 1973.
113. **Goldstein, J. A., Reller, L. B., and Wang, W.-L. L.,** Limulus amebocyte lysate test in neonates, *Am. J. Clin. Pathol.,* 66, 1012, 1976.
114. **Levin, J., Poore, T. E., Zauber, N. P., and Oser, R. S.,** Detection of endotoxin in the blood of patients with sepsis due to Gram-negative bacteria, *N. Engl. J. Med.,* 283, 1313, 1970.
115. **Levin, J., Poore, T. E., Young, N. S., Margolis, S., Zauber, N. P., Townes, A. S., and Bell, W. R.,** Gram-negative sepsis: detection of endotoxemia with the limulus test, *Ann. Intern. Med.,* 76, 1, 1972.
116. **Oberle, M. W., Graham, G. G., and Levin, J.,** Detection of endotoxemia with the Limulus test: preliminary studies in severely malnourished children, *J. Pediatr.,* 85, 570, 1974.
117. **Caridis, D. T., Reinhold, R. B., Woodruff, P. W. H., and Fine, J.,** Endotoxaemia in man, *Lancet,* 1, 1381, 1972.
118. **Levin, J.,** Endotoxin and endotoxemia, *N. Engl. J. Med.,* 288, 1297, 1973.
119. **Martinez-G, L. A., Quintiliani, R., and Tilton, R. C.,** Clinical experience on the detection of endotoxemia with the Limulus test, *J. Infect. Dis.,* 127, 102, 1973.

120. Stumacher, R. J., Kovnat, M. J., and McCabe, W. R., Limitations of the usefulness of the limulus assay for endotoxin, *N. Engl. J. Med.*, 288, 1261, 1973.
121. Feldman, S. and Pearson, T. A., The *Limulus* test and Gram-negative bacillary sepsis, *Am. J. Dis. Child.*, 128, 172, 1974.
122. Elin, R. J., Robinson, R. A., Levine, A. S., and Wolff, S. M., Letters to the editor, *N. Engl. J. Med.*, 294, 48, 1976.
123. Berman, N. S., Siegel, S. E., Nachum, R., Lipsey, A., and Leedom, J., Cerebrospinal fluid endotoxin concentrations in Gram-negative bacterial meningitis, *J. Pediatr.*, 88, 553, 1976.
124. McCracken, G. H., Jr. and Sarff, L. D., Endotoxin in cerebrospinal fluid. Detection in neonates with bacterial meningitis, *JAMA*, 235, 617, 1976.
125. Nachum, R., Lipsey, A., Siegel, S. E., Rapid detection of Gram-negative bacterial meningitis by the limulus lysate test, *N. Engl. J. Med.*, 289, 931, 1973.
126. Ross, S., Rodriguez, W., Controni, G., Korengold, G., Watson, S., and Khan, W., Limulus lysate test for Gram-negative bacterial meningitis, *JAMA*, 233, 1366, 1975.
127. Jorgensen, J. H., Carvajal, H. F., Chipps, B. E., and Smith, R. F., Rapid detection of Gram-negative bacteriuria by use of the *Limulus* endotoxin assay, *Appl. Microbiol.*, 26, 38, 1973.
128. Abel, K., deSchmertzing, H., and Peterson, J. I., Classification of microorganisms by analysis of chemical composition. I. Feasibility of utilizing gas chromatography, *J. Bacteriol.*, 85, 1039, 1963.
129. Bricknell, K. S., Sutter, V. L., and Finegold, S. M., Detection and identification of anaerobic bacteria, in *Gas Chromatographic Applications in Microbiology and Medicine,* Mitruka, B. M., Ed., John Wiley & Sons, New York, 1975.
130. Cherry, W. B. and Moss, C. W., The role of gas chromatography in the clinical microbiology laboratory, *J. Infect. Dis.*, 119, 658, 1969.
131. Mitruka, B. M., Ed., *Gas Chromatographic Applications in Microbiology and Medicine,* John Wiley & Sons, New York, 1975.
132. Moss, C. W., Gas-liquid chromatography as an analytical tool in microbiology, *Public Health Lab.*, 33, 81, 1975.
134. Mitruka, B. M., Jonas, A. M., and Alexander, M., Rapid detection of bacteremia in mice by chromatography, *Infect. Immun.*, 2, 474, 1970.
135. Mitruka, B. M., Kundargi, R. S., and Jonas, A. M., Gas chromatography for rapid differentiation of bacterial infections in man, *Med. Res. Eng.*, 11, 7, 1972.
136. Mitruka, B. M., Rapid automated identification of microorganisms in clinical specimens by gas chromatography, in *New Approaches to the Identification of Microorganisms,* Heden, C.-G. and Illeni, T., Eds., John Wiley & Sons, New York, 1975.
137. Davis, C. E. and McPherson, R. A., Rapid diagnosis of septicemia and meningitis by gas-liquid chromatography, in *Microbiology–1975,* Schlessinger, D., Ed., American Society for Microbiology, Washington, D.C., 1975.
138. Brooks, J. B., Cherry, W. B., Thacker, L., and Alley, C. C., Analysis by gas chromatography of amines and nitrosamines produced in vivo and in vitro by *Proteus mirabilis, J. Infect. Dis.*, 126, 143, 1972.
139. Brooks, J. B., Kellogg, D. S., Alley, C. C., Short, H. B., Handsfield, H. H., and Huff, B., Gas chromatography as a potential means of diagnosing arthritis. I. Differentiation between staphylococcal, streptococcal, gonococcal, and traumatic arthritis, *J. Infect. Dis.*, 129, 660, 1974.
140. Controni, G., Rodriguez, W., Deane, C., Ross, S., Puig, J., and Khan, W., Rapid diagnosis of meningitis by gas liquid chromatographic analysis of cerebrospinal fluid lactic acid, *Abstr. Ann. Meeting ASM,* C23, 1975.
141. Amundson, S., Braude, A. I., and Davis, C. E., Rapid diagnosis of infection by gas-liquid chromatography: analysis of sugars in normal and infected cerebrospinal fluid, *Appl. Microbiol.*, 28, 298, 1974.
142. Gorbach, S. L., Mayhew, J. W., Bartlett, J. G., Thadepalli, H., and Onderdonk, A. B., Rapid diagnosis of anaerobic infections by direct gas-liquid chromatography of clinical specimens, *J. Clin. Invest.*, 57, 478, 1976.
143. Maccani, J. E. and Finn, G. R., Sr., Experience using direct gas-liquid chromatography of clinical material to identify anaerobic infection, *Abstr. Ann. Meeting ASM,* C141, 1977.
144. Phillips, K. D., Tearle, P. V., and Willis, A. T., Rapid diagnosis of anaerobic infections by gas-liquid chromatography of clinical material, *J. Clin. Pathol.*, 29, 428, 1976.
145. Wadstrom, T. and Nord, C.-E., Immunoelectroosmophoresis and electroimmunoassays in clinical microbiology, in *Automation in Microbiology and Immunology,* Heden, C.-G. and Illeni, T., Eds., John Wiley & Sons, New York, 1975.
146. Rytel, M. W., Rapid diagnostic methods in infectious diseases, *Adv. Intern. Med.*, 20, 37, 1975.
147. Verbruggen, R., Quantitative immunoelectrophoretic methods: a literature survey, *Clin. Chem.*, 21, 5, 1975.
148. Anhalt, J. P. and Wee, S. H., Counterimmunoelectrophoresis of pneumococcal antigens: alternative buffers for the detection of types VII and XIV, *Abstr. 16th Intersci. Conf. Antimicrob. Agents Chemother.*, 312, 1976.
149. Kenny, G. E., Wentworth, B. B., Beasley, R. P., and Foy, H. M., Correlation of circulating capsular polysaccharide with bacteremia in pneumococcal pneumonia, *Infect. Immun.*, 6, 431, 1972.
150. Anhalt, J. P. and Yu, P. K. W., Counterimmunoelectrophoresis of pneumococcal antigens: improved sensitivity for the detection of types VII and XIV, *J. Clin. Microbiol.*, 2, 510, 1975.
151. Coonrod, J. D. and Rytel, M. W., Determination of aetiology of bacterial meningitis by counterimmunoelectrophoresis, *Lancet,* 1, 1154, 1972.

152. **Edwards, E. A., Muehl, P. M., and Peckinpaugh, R. O.,** Diagnosis of bacterial meningitis by counterimmunoelectrophoresis, *J. Lab. Clin. Med.,* 80, 449, 1972.
153. **Feigin, R. D., Stechenberg, B. W., Chang, M. J., Dunkle, L. M., Wong, M. L., Palkes, H., Dodge, P. R., and Davis, H.,** Prospective evaluation of treatment of *Haemophilus influenzae* meningitis, *J. Pediatr.,* 88, 542, 1976.
154. **Feigin, R. D., Wong, M., Shackelford, P. G., Stechenberg, B. W., Dunkle, L. M., and Kaplan, S.,** Countercurrent immunoelectrophoresis of urine as well as of CSF and blood for diagnosis of bacterial meningitis, *J. Pediatr.,* 89, 773, 1976.
155. **Ingram, D. L., Anderson, P., and Smith, D. H.,** Countercurrent immunoelectrophoresis in the diagnosis of systemic diseases caused by *Hemophilus influenzae* type b, *J. Pediatr.,* 81, 1156, 1972.
156. **Ingram, D. L., O'Reilly, R. J., Peter, G., Anderson, P., and Smith, H.,** Systemic capsular antigen, clinical course and antibody response in *Hemophilus influenzae* b meningitis (abstract), *Pediatr. Res.,* 7, 369, 1973.
157. **Shackelford, P. G., Campbell, J., and Feigin, R. D.,** Countercurrent immunoelectrophoresis in the evaluation of childhood infections, *J. Pediatr.,* 85, 478, 1974.
158. **Smith, E. W. P. and Ingram, D. L.,** Counterimmunoelectrophoresis in *Hemophilus influenzae* type b epiglottitis and pericarditis, *J. Pediatr.,* 86, 571, 1975.
159. **Fossieck, B., Jr., Craig, R., and Paterson, P. Y.,** Counterimmunoelectrophoresis for rapid diagnosis of meningitis due to *Diplococcus pneumoniae, J. Infect. Dis.,* 127, 106, 1973.
160. **Bartram, C. E., Jr., Crowder, J. G., Beeler, B., and White, A.,** Diagnosis of bacterial diseases by detection of serum antigens by counterimmunoelectrophoresis, sensitivity, and specificity of detecting *Pseudomonas* and pneumococcal antigens, *J. Lab. Clin. Med.,* 83, 591, 1974.
161. **Coonrod, J. D. and Rytel, M. W.,** Detection of type-specific pneumococcal antigens by counterimmunoelectrophoresis. II. Etiologic diagnosis of pneumococcal pneumonia, *J. Lab. Clin. Med.,* 81, 778, 1973.
162. **Coonrod, J. D.,** Physical and immunologic properties of pneumococcal capsular polysaccharide produced during human infection, *J. Immunol.,* 112, 2193, 1974.
163. **Coonrod, J. D. and Drennan, D. P.,** Pneumococcal pneumonia: capsular polysaccharide antigenemia and antibody responses, *Ann. Intern. Med.,* 84, 254, 1976.
164. **Dorff, G. J., Coonrod, J. D., and Rytel, M. W.,** Detection by immunoelectrophoresis of antigen in sera of patients with pneumococcal bacteremia, *Lancet,* 1, 578, 1971.
165. **Perlino, C. A. and Shulman, J. A.,** Detection of pneumococcal polysaccharide in the sputum of patients with pneumococcal pneumonia by counterimmunoelectrophoresis, *J. Lab. Clin. Med.,* 87, 496, 1976.
166. **Spencer, R. C. and Savage, M. A.,** Use of counter and rocket immunoelectrophoresis in acute respiratory infections due to *Streptococcus pneumoniae, J. Clin. Pathol.,* 29, 187, 1976.
167. **Tugwell, P. and Greenwood, B. M.,** Pneumococcal antigen in lobar pneumonia, *J. Clin. Pathol.,* 28, 118, 1975.
168. **Foy, H. M., Wentworth, B., Kenny, G. E., Kloeck, J. M., and Grayston, J. T.,** Pneumococcal isolations from patients with pneumonia and control subjects in a prepaid medical care group, *Am. Rev. Respir. Dis.,* 111, 595, 1975.
169. **Rytel, M. W., Dee, T. H., Ferstenfeld, J. E., and Hensley, G. T.,** Possible pathogenetic role of capsular antigens in fulminant pneumococcal disease with disseminated intravascular coagulation, *Am. J. Med.,* 57, 889, 1974.
170. **Edwards, E. A.,** Immunologic investigations of meningococcal disease. I. Group-specific *Neisseria meningitidis* antigens present in the serum of patients with fulminant meningococcemia, *J. Immunol.,* 106, 314, 1971.
171. **Greenwood, B. M., Whittle, H. C., and Dominic-Rajkovic, O.,** Counter-current immunoelectrophoresis in the diagnosis of meningococcal infections, *Lancet,* 2, 519, 1971.
172. **Hoffman, T. A. and Edwards, E. A.,** Group-specific polysaccharide antigen and humoral antibody response in disease due to *Neisseria meningitidis, J. Infect. Dis.,* 126, 636, 1972.
173. **Whittle, H. C., Greenwood, B. M., Davidson, N. McD., Tomkins, A., Tugwell, P., Warrell, D. A., Zalin, A., Bryceson, A. D. M., Parry, E. H. O., Brueton, M., Duggan, M., Oomen, J. M. V., and Rajkovic, A. D.,** Meningococcal antigen in diagnosis and treatment of group A meningococcal infections, *Am. J. Med.,* 58, 823, 1975.
174. **Tobin, B. M. and Jones, D. M.,** Immunoelectroosmophoresis in the diagnosis of meningococcal infections, *J. Clin. Pathol.,* 25, 583, 1972.
175. **McCracken, G. H., Jr., Sarff, L. D., Glode, M. P., Mize, S. G., Schiffer, M. S., Robbins, J. B., Gotschlich, E. C., Orskov, I., and Orskov, F.,** Relation between *Escherichia coli* K1 capsular polysaccharide antigen and clinical outcome in neonatal meningitis, *Lancet,* 2, 246, 1974.
176. **Jackson, L. J., Aguilar-Torres, F. G., Dorado, A., Rose, H. D., and Rytel, M. W.,** Detection of staphylococcal teichoic acid antigen in body fluids and tissue, *Abstr. Ann. Meeting ASM,* C5, 1977.
177. **Lampe, R. M., Chottipitayasunondh, T., and Sunakorn, P.,** Detection of bacterial antigen in pleural fluid by counterimmunoelectrophoresis, *J. Pediatr.,* 88, 557, 1976.
178. **Hill, H. R., Riter, M. E., Menge, S. K., Johnson, D. R., and Matsen, J. M.,** Rapid identification of group B streptococci by counterimmunoelectrophoresis, *J. Clin. Microbiol.,* 1, 188, 1975.
179. **Hill, H. R.,** Rapid detection and specific identification of infections due to group B streptococci by counterimmunoelectrophoresis, in *Microbiology – 1975,* Schlessinger, D., Ed., American Society for Microbiology, Washington, D.C., 1975.
180. **Adams, M. B., Aguilar-Torres, F. G., and Rytel, M. W.,** Detection of *Pseudomonas* antigen by counterimmunoelectrophoresis, *Surg. Forum,* 27, 3, 1976.

181. **Simpson, R. A. and Speller, D. C. E.,** Detection of bacteraemia by countercurrent immunoelectrophoresis, *Lancet,* 1, 1206, 1977.
182. **Coonrod, J. D. and Rytel, M. W.,** Specificity of counterimmunoelectrophoresis in bacterial meningitis, *Lancet,* 2, 829, 1972.
183. **Ribner, B., Keusch, G. T., and Robbins, J. B.,** *Staphylococcus aureus* antigen in cerebrospinal fluid cross-reactive with *Haemophilus influenzae* type b antiserum, *Ann. Intern. Med.,* 83, 370, 1975.
184. **Colding, H. and Lind, I.,** Counterimmunoelectrophoresis in the diagnosis of bacterial meningitis, *J. Clin. Microbiol.,* 5, 405, 1977.
185. **Waters, J. R.,** personal communication, Johnston Laboratories, Cockeysville, MD, 1977.
186. **Zierdt, C. H.,** personal communication, Clinical Pathology Department, Clinical Center, National Institutes of Health, Bethesda, Md., 1977.

APPENDIX

RECOMMENDED PROCEDURES FOR CULTURE OF BLOOD FOR BACTERIA

I. Blood collection

A. Guidelines for the number of and intervals between blood collections according to the type of disease present are as follows:

Infection	Bacteremia	Collection	
		Number[a]/24 hr	Interval[b]
Endocarditis, early typhoid fever or brucellosis, severe uncontrolled infection	Continuous	2–3	Any
Other septicemias	Intermittent	3	1 hr

[a]A larger number may be indicated when patient has recently received or is receiving antimicrobial agents.
[b]The interval should be shortened if antimicrobial therapy is to be initiated shortly.

B. Suggested volumes of blood per collection are as follows:

	ml
Adults	20–30
Infants and small children	1–3

C. Skin antisepsis – Careful antisepsis is essential to minimize contamination of cultures. An alcohol (70 to 95%) scrub followed by an iodine (2%) or iodophor preparation of the venipuncture site is satisfactory.

II. Media

A. Composition – The selection of media is complicated by differences among the same generic products made by various manufacturers; therefore, the performance of one manufacturer's product, e.g., brain heart infusion broth, may differ from that made by another manufacturer. Studies at the Mayo Clinic with one manufacturer's blood culture bottles (Difco Laboratories, Detroit, Mich.) have shown equivalent performances by soybean casein digest (Tryptic soy), brain heart infusion, and Brucella broths. Thioglycollate and thiol (Difco Laboratories, Detroit, Mich.) have been shown to be equivalent in performance to soybean casein digest in every respect except for their significantly lower yield of pseudomonads and yeasts. Columbia broth has been shown to be equivalent to soybean casein digest broth in every respect except for its significantly lower yield of *Staphylococcus aureus.*

Soybean casein digest broth is, therefore, recommended for routine use, although consumers should be aware of the fact that the various manufacturers' formulations of this product may differ and that equivalency in performance cannot be taken for granted. The substitution of soybean casein digest broth in one of a two- or three-bottle system with another medium is acceptable though probably unnecessary.

B. Volume – Until clinical studies demonstrate that a minimum 10% v/v dilution of blood in broth is unnecessary, this requirement will govern the volume of broth inoculated. It is, therefore, recommended that 20 or 30 ml of blood from adults be distributed equally into each of two or three 100-ml bottles.

C. Type

1. Osmolarity – The value of media rendered hyperosmolar, usually with added sucrose, remains unclear, but it is most likely both medium and system dependent.
2. Prereduced anaerobically sterilized (PRAS) – The theoretical advantages of such media for recovering anaerobes have failed to be substantiated in most clinical studies comparing PRAS media to those containing ordinary media in vacuum bottles with CO_2.

D. Inactivators

1. Sodium polyanetholsulfonate (SPS, Liquoid®) – SPS significantly increases the recovery of Gram-positive and Gram-negative bacteria from blood and should be included routinely in concentrations of 0.025 to 0.05% in blood culture media. One notable exception to this recommendation is in cultures for *Neisseria gonorrhoeae* and *Neisseria meningitidis* for which SPS is inhibitory. The inhibitory effects of SPS on *Peptostreptococcus anaerobius* are irrelevant to blood cultures.
2. Sodium amylosulfate (SAS) – Similar in activity to SPS but inactive against *P. anaerobius,* SAS has been shown to be inhibitory to *S. aureus* and to reduce the rate of isolation of Gram-negative bacilli when compared to SPS. Its use in blood culture media is contraindicated.
3. Penicillinase – Although this method has not been critically evaluated for many years, it is suggested that penicillinase be added to media inoculated with blood from patients who have recently or are receiving penicillins. It is strongly recommended that it be added only to one bottle per set and that its sterility be confirmed concurrently.

III. Incubation

A. Temperature – It is suggested that incubators be adjusted to 35°C.

B. Atmosphere – When using vacuum blood culture bottles, it is necessary to inoculate at least two bottles and to vent one transiently prior to incubation.

C. Duration – Bottles should be incubated for a minimum of 5 days. In the majority of instances, 7 days are satisfactory. Longer periods may be necessary in cases of suspected endocarditis or uncontrolled sepsis with negative cultures after 7 days' incubation.

IV. Examination

A. Macroscopic – Bottles should be inspected later in the day of their receipt in the laboratory and daily thereafter.

B. Microscopic – Any medium suspected of containing bacteria on the basis of macroscopic examination should be smeared and stained immediately. The value of routine Gram's-stained smears of macroscopically negative bottles is controversial; however, such examinations are probably unnecessary if early subcultures are routinely performed (see V, B below).

V. Subculture

A. Suspected positive – Subcultures of media suspected of being positive should be made into media suitable for the growth of both aerobic and anaerobic bacteria.

B. Routine ("blind") – Routine subcultures of macroscopically negative media are mandatory. The optimal time for subculture is on the day the medium was inoculated with blood. Early subcultures should be made routinely of all bottles between 1 and 14 hr following their inoculation. An additional routine subculture may be made after 48 to 72 hr of incubation; further routine subcultures are unnecessary.

Subcultures should be made by aspiration of an aliquot of the blood-broth mixture with a sterile needle and syringe. This aliquot is inoculated onto a properly labeled quadrant of a chocolate blood agar plate that is incubated for 48 hr in an atmosphere containing 5 to 10% CO_2. Routine anaerobic subcultures are unnecessary.

VI. Significance of positive cultures

A. Presumed contaminants – *Bacillus, Corynebacterium, Propionibacterium,* and *S. epidermidis* usually represent contaminants unless isolated from multiple cultures. Interpretation of their significance is greatly facilitated when a minimum of two separate sets of blood cultures is required. Their isolation should nonetheless be reported.

B. Nearly always clinically significant – β-hemolytic streptococci, *Streptococcus pneumoniae, Escherichia coli, Salmonella,* Klebsielleae, Proteeae, Bacteroidaceae, haemophili, *Pseudomonas aeruginosa, Staphylococcus aureus,* and clostridia are nearly always clinically significant when isolated from blood.

C. Uncertain clinical significance – Viridans and group D streptococci are of questionable significance unless isolated from multiple sets of blood cultures.

VII. Supplementary procedures

A. Identification and antimicrobial susceptibility testing – In many instances, direct inoculation of biochemical tests can be made with a broth or washed suspension of organisms from a positive blood culture. In nearly all instances, with the exception of polymicrobial bacteremias, direct inoculation of a standardized inoculum can be made for antimicrobial susceptibility testing.

B. Pour plates – The value of pour plates remains controversial. They are somewhat less sensitive than broth cultures, especially in detecting bacteremias of a low order of magnitude. Although generally considered to be unnecessary, some investigators have found them to be helpful in distinguishing between contaminated and uncontaminated cultures; therefore, they may be of value in hospitals where contamination of blood cultures represents a major problem.

C. Cell wall defective (CWD) bacterial variants ("L forms") – CWD variants have been isolated rarely from blood but are not to be confused with isolates from hyperosmolar media. Their isolation requires careful and scrupulously controlled techniques, is very time consuming, and should, therefore, be reserved for highly selected cases and research laboratories.

D. *Brucella* – Brucellae are optimally isolated by prolonged (30 day minimum) incubation of blood in a bottle containing biphasic soybean casein digest medium and 10% CO_2.

E. *Leptospira* – Cultures of blood for leptospires should be restricted to the first week of acute illness. A few (one to three) drops of blood should be added to each of four tubes with semisolid medium which are then incubated at 30°C for 28 days.

VIII. Reports – The patient's attending physician must be notified by phone and in writing of any findings in blood cultures. The initial report ordinarily consists of the Gram's-stained smear findings but should also reflect the number of cultures and bottles in which the organism is present. Additional reports consist of the organism's identification and its susceptibility to various antimicrobial agents. All telephoned reports must be confirmed in writing, and all written reports should bear a notation of the date and time of the telephoned report and the initials of the person who made the call.

INDEX

A

B

C

D

E

F

H

I

K

L

M

N

P

S

T

U

V

CRC PUBLICATIONS OF RELATED INTEREST

CRC HANDBOOKS:

CRC HANDBOOK OF MICROBIOLOGY, 2nd Edition
Edited by **Allen I. Laskin, Ph.D.,** Exxon Research and Engineering Company, and **Hubert Lechevalier, Ph.D.,** Rutgers University.
Current, comprehensive information in tabular form and text, on the properties of microorganisms, their composition, products, and activities.

CRC HANDBOOK SERIES IN CLINICAL LABORATORY SCIENCE
Editor-in-Chief, **David Seligson, Sc.D., M.D.,** Yale University School of Medicine.
This unique Series offers a comprehensive reference source for selected areas of clinical laboratory science: Blood Banking, Clinical Chemistry, Hematology, Immunology, Microbiology, Pathology, Virology, Toxicology, and Nuclear Medicine.

CRC HANDBOOK OF LABORATORY ANIMAL SCIENCE
Edited by **Edward C. Melby, Jr., D.V.M.,** New York State College of Veterinary Medicine, and **Norman H. Altman, V.M.D.,** Papanicolaou Cancer Research Institute.
A multi-volume reference source of information covering all aspects of the selection and use of laboratory animals for teaching, research, and testing.

CRC HANDBOOK OF CHROMATOGRAPHY
Edited by **Gunter Zweig, Ph.D.,** Environmental Protection Agency, and **Joseph Sherma, Ph.D.,** Lafayette College.
A two-volume presentation of comprehensive data, methods, and literature, including more than 12,000 conveniently listed compounds, designed to provide a working knowledge of the theory and practices of chromatography: gas, liquid column, paper, and thin-layer.

CRC COMPOSITE INDEX, 2nd Edition
Provides convenient access to specific data contained in 49 current CRC Handbooks, including a special chemical substances section and more than 250,000 entries.

CRC HANDBOOK OF CLINICAL LABORATORY DATA, 2nd Edition
Edited by **Willard R. Faulkner, Ph.D.,** Vanderbilt University, **John W. King, M.D., Ph.D.,** Cleveland Clinic Foundation, and **Henry C. Damm, Ph.D.,** Cleveland State University.
Every phase of the modern laboratory is covered for the microbiologst, clinical chemist, pathologist, and hematologist.

CRC MANUAL OF CLINICAL LABORATORY PROCEDURES, 2nd Edition
Edited by **Willard R. Faulkner, Ph.D.,** Vanderbilt University, and **John W. King, M.D., Ph.D.,** Cleveland Clinic Foundation.
A practical workbook describing principles, normal range of results, equipment requirements, sampling procedures, reagents, procedural instructions, and calculation of results.

CRC UNISCIENCE PUBLICATIONS:

METHODS OF DETECTION AND IDENTIFICATION OF BACTERIA
By **Brij M. Mitruka, M.S., Ph.D.,** and **Mary J. Bonner, M.S., Ph.D.,** both with the University of Pennsylvania Medical Center.
This book encompasses conventional methods and recent developments in the methodology of detection and identification of bacteria in *in vitro* cultures and biological samples.

URINALYSIS IN CLINICAL LABORATORY PRACTICE
By **Alfred H. Free, Ph.D.,** and **Helen M. Free,** both with the Ames Division of Miles Laboratories.
A wide range of topics is covered: from routine screening to esoteric considerations used only in rare instances; handling practices; specific urine constituents; parameters of measurement; and urine analysis with regard to diabetics and pediatric patients.

CELL WALL DEFICIENT FORMS
By **Lida H. Mattman, Ph.D.,** Wayne State University.
This book signifies the importance of studying microorganisms defective in cell wall synthesis, and which play an outstanding role in tuberculosis, acute infections, and other clinical conditions.

DECISION ANALYSIS IN MEDICINE: METHODS AND APPLICATIONS
By **Edward A. Patrick, M.D., Ph.D.,** Purdue University.
Statistical pattern recognition is applied to medicine, using a model of diagnosis and consulting, to follow the changing condition of the patient, in addition to cost-benefit ratios, and automated prognosis systems.

CHEMOTHERAPY OF INFECTIOUS DISEASE
By **Hans H. Gadebusch, Ph.D.,** Squibb Institute for Medical Research.
Models of animal disease that closely mimic infection in humans are discussed with relation to their effect on studies with bacteria, yeasts, fungi, protozoa, helminths, and viruses.

CRC CRITICAL REVIEW JOURNALS:

CRC CRITICAL REVIEWS™ IN MICROBIOLOGY
Edited by **Allen I. Laskin, Ph.D.,** Exxon Research and Engineering Co., and **Hubert Lechevalier, Ph.D.,** Rutgers University.

CRC CRITICAL REVIEWS™ IN CLINICAL LABORATORY SCIENCES
Edited by **John Batsakis, M.D.,** University of Michigan Medical School, and **John Savory, Ph.D.,** University of North Carolina.

Please forward inquiries to CRC Press, Inc., 2255 Palm Beach Lakes Boulevard, West Palm Beach, Florida 33409 U.S.A.

RC
182
.S4
D39